Tales from Toothaker

How We Used Humor, Hard Work, and Hand-Me-Downs to Create an Island Home

Peggy Herbert

iUniverse, Inc.
Bloomington

Tales from Toothaker
How We Used Humor, Hard Work, and Hand-Me-Downs to Create an Island Home

Copyright © 2011 by Peggy Herbert

All rights reserved. No part of this book may be used or reproduced by any means, graphic, electronic, or mechanical, including photocopying, recording, taping or by any information storage retrieval system without the written permission of the publisher except in the case of brief quotations embodied in critical articles and reviews.

iUniverse books may be ordered through booksellers or by contacting:

iUniverse
1663 Liberty Drive
Bloomington, IN 47403
www.iuniverse.com
1-800-Authors (1-800-288-4677)

Because of the dynamic nature of the Internet, any web addresses or links contained in this book may have changed since publication and may no longer be valid. The views expressed in this work are solely those of the author and do not necessarily reflect the views of the publisher, and the publisher hereby disclaims any responsibility for them.

Any people depicted in stock imagery provided by Thinkstock are models, and such images are being used for illustrative purposes only.

Certain stock imagery © Thinkstock.

ISBN: 978-1-4620-0743-1 (sc)
ISBN: 978-1-4620-0741-7 (ebook)
ISBN: 978-1-4620-0742-4 (dj)

Library of Congress Control Number: 2011904554

Printed in the United States of America

iUniverse rev. date: 4/18/2011

For Tom---who made it all possible

Acknowledgements

I'D LIKE TO EXPRESS MY appreciation to all of the students that I taught over the years in Henniker. While my job was to inspire them, they also inspired me. They were my first audience as I wrote stories about the island to model the writing process. Their feedback and enthusiasm kept me writing, first for them, and then just for fun.

I'd also like to thank our island neighbor Chris Hastedt, who upon reading early drafts of Tales From Toothaker became adamant that I publish it. When my enthusiasm flagged, she gave me the necessary prodding to keep the process moving ahead.

There are a few people whose help over the years has made the cabin a reality. Holly and Lenny Charron, LeRoy and Diana Anderson, Brenda Fortier-Dube, and Tom LaJoie all have consistently been there to make what started as a vision, into a reality. They have also provided me with a lot of great stories!

To my beautiful daughters, Kelly and Lindsey, my wonderful son-in-law Todd and to our three grandchildren, Emma, Cooper and Regan; your love of the island fulfills what your father and I hoped for all those years ago when we envisioned the future.

And finally, to Tom whose endless energy and boundless creativity has made the cabin and therefore this book possible.

Contents

1 ~ How It Came to Be — 1
2 ~ A Little Bit About Us — 5
3 ~ A History of the Mooselookmeguntic Area — 9
4 ~ A Tent is Not Enough — 17
5 ~ A Simple Cabin is Not Enough — 23
6 ~ Outfitting Our *Simple* Cabin — 31
7 ~ Cooking on Toothaker — 39
8 ~ What We Do When There's Nowhere to Go — 45
9 ~ Tom's Trails — 51
10 ~ Boats….An Island Necessity — 57
11 ~ Ever-Evolving Dock Systems — 67
12 ~ Our "Neighborhood" — 73
13 ~ Guests, Expected and the Unexpected — 77
14 ~ What to Be Afraid Of — 81
15 ~ Island Critters — 91
16 ~ Kayaking Richardson — 95
17 ~ Having Fun at 20 Below — 97
18 ~ The Next Generation — 103
19 ~ Stories from the Grandchildren — 109
20 ~ The End…or Just More Beginnings? — 113

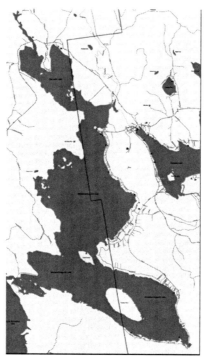

Mooselookmeguntic Lake (photo from Rangeley-Maine.com and RangeleyVacations)

1 ~ How It Came to Be

In the late summer of 1985, my husband Tom and I returned to our central New Hampshire home from a three day camping trip in Maine with dirty clothes, an empty cooler and a commitment to buy fifty acres of land on Toothaker island in Lake Mooselookmeguntic. We had not gone to Maine looking to buy anything. Even when we went into the realtor's office we were only going out of curiosity.

When the realtor told us of two lots on Toothaker Island, a seven hundred fifty acre island on the south end of Mooselookmeguntic, we were intrigued. We had always dreamed of owning a cabin on a lake. Tom's brother Peter has had a water accessible place in northern Minnesota since the mid-sixties, so the idea of not being able to drive to a cabin did not seem strange. We got directions and canoed over to see if we could locate the pieces that were for sale.

We never found one of them, but the one we did find faced south, had four hundred feet of shoreline, bordered on preserved land and had a good place to put a cabin. The fact that we couldn't get more than ten feet on shore because of the blow-downs and undergrowth did not deter us. We saw a vision of what was to be, never dreaming for a minute of the challenges we would face getting there.

Lake Mooselookmeguntic is the largest of a chain of lakes proceeding from Rangeley Lake, to Mooselook and Cupsuptic, into Upper and Lower Richardson, which are then connected to Lake Umbagog by way of the Rapid River. Umbagog empties into the Androscoggin River, which makes its way to Berlin, New Hampshire and on to Rumford, Maine; both, at some point, homes to large paper mills. In the early logging days, the lakes were part of the log highway, used to transport the timber to the mills. Spruce forests surround the lake and the log highway is still with us but in the form of huge lumber trucks that we hear going back and forth on the dirt roads that web the hills.

The island, named for an early Rangeley settler, sits about a half mile out from the south shore of the lake. The land directly across from us and down to the west all belongs to the state. Even those cabins further up to the east are hard to see, as they are all set back from the shore. Behind rise low mountains that are part of the Appalachian Trail. The lake is shaped like a boot and the island is in the foot of the boot. At the heel there is a dam that controls the level of the lake. In May, the water level is high and we have no beach. During the summer, the water drops until by late fall, the water's edge is about twenty-five feet from the high water mark.

Holly walking on the beach in the fall. In the spring, the large rock on the right would be totally under water.

When we bought our fifty acres, there were no other cabins on the south side of the island and only two on the north side--one of them inhabited by a couple who lived there year round. Now twenty-five years later, there are ten cabins on one side and twelve on the other. But back then it was a very desolate place. The nearest landing (Indian Cove) was a mile away and that was not truly a landing--more of a beach that allowed us to launch our canoes. When we had a motorboat and could no longer use Indian Cove, we had to travel three miles across mostly open water from Bemis Landing, an official landing with a ramp that we had access to through the realtor.

One of the first things we learned is the importance of the weather not only in the mountains of western Maine but on the lake itself. Summers are generally lovely with temperatures ranging anywhere from fifty to ninety degrees. There can be days of rain or long periods of bright sunshine.

Winters are cold--the ice is often thirty inches thick and Tom has camped up here when it was twenty below zero. The area gets lots of snow. When we have eight inches of snow on the ground in New Hampshire, there will be eighteen inches of snow on Toothaker.

But the one constant aspect of the weather in both winter and summer is the wind. It is a rare day when the wind does not start blowing, usually about 10 am. The pine trees wave their branches like gentle giants…and if the wind is strong enough, their shallow root systems will rise and fall under the ground looking like something is breathing just beneath the surface. Each spring we arrive wondering which trees will have blown over during the winter storms When we first built the cabin, we left four very tall pines standing on the west side. When we came back in the spring, we discovered one of the trees leaning against the cabin wall after chewing off a piece of our roof on its way down. We quickly cut down the remaining three to avoid future problems.

In the fall, we often wake up to dense fog that obliterates everything but what is right in front of us. The world seems to end just beyond our shore and we wait, encased in a cocoon, for the sun to burn away our misty shell, which sometimes doesn't happen until ten or eleven. We did get a compass after failing to find the boat launch one morning and now have a bearing to follow in the event we need to need to get off the island during one of these foggy mornings.

Sometimes the weather is just plain weird. I remember waking early one morning in July to rumbling overhead. My first thought was thunder, but as the noise went steadily on down the lake, I thought it was the sound of an airplane. Just as it began to fade, I heard it again. I imagined it circling, but that made no sense, as jets do not circle Lake Mooselookmeguntic. I began to question my assumption that it was an airplane.

When I finally opened my eyes and discovered a most peculiar yellow tinge to the atmosphere, I realized that it was indeed thunder. Everything was deadly still and the trees looked surrealistic as they stood there in the yellow haze. It was not like sunlight and it was certainly unlike the predawn gray that we normally awaken to.

So I lay there listening, watching, waiting…for I was sure something dramatic was going to happen. Then, as though someone had flipped a switch, the yellow was gone and the air was its usual morning gray. The thunder faded and it started to rain, steadily, drearily. I felt disappointed. Such an unusual beginning should have materialized into more than another wet day at the lake.

One is always more aware of the weather here on the island because so much depends on it. If the water is too rough, it is hard to get on or off the island. If it is too cold, the woodstove has to be started. If it is going to storm, one hopes not to be somewhere out on the lake in a canoe with rough water and lightning.

But of course, we didn't know all of this when we agreed to buy our land. We just knew we'd found a place to realize our dream of a cabin on a lake in the northern forests of New England.

2 ~ A Little Bit About Us

When we bought our island land, Tom was a forty-two year old high school social studies teacher in Concord, New Hampshire. He had been born in New Jersey and had spent many summers at a small boys' camp in southern Maine.

He was the more adventuresome of the two of us. He had been running a modified Outward Bound program at his school, and his motto was *plus est en vous,* which translates to "more is in you." He thought nothing of taking twenty high school students out for a three-day backpacking trip in January. Or bicycling two hundred miles in June. His goal was always getting his students to realize that they could do more than they ever imagined, a lesson they could then carry with them as they faced struggles in their lives.

At home, having had very little carpentry experience, he decided to build a woodshed. While it had an unusual slant, it didn't stop him from adding a family room onto our hundred-year-old house or tearing out our kitchen and installing a new one. He had the can-do attitude that we'd need for this project.

I had been born in Minnesota, the land of 10,000 lakes, and while we never owned a cabin, our family spent a lot of time enjoying the water. As the oldest of six children, I had a tendency to bossiness…in fact I much preferred telling people what to do than doing it myself. Tom was the youngest of four and was a perfect match for me in that he was used to taking orders. Or so I thought. He'd listen, nod his head and then go and do what he wanted. Like Tom, I was a teacher as well, and so we almost always have had our summers free to be at the island.

We have two daughters, Kelly and Lindsey. When we first bought the land, Kelly was fourteen, an age where spending time at an island away from her friends was the last thing she wanted to do. However, now that she is in her thirties with her own three kids, she and her husband Todd Anderson are as passionate and committed to the cabin as we are. Kelly is the person who coined the term "awkward manual labor" (versus just the regular old manual labor). Unfortunately most of our manual labor here on the island tends to be of the awkward variety. She is also the first person to have spent a month on the island, complete with an infant and two toddlers, never leaving except for a run into Rangeley.

Lindsey is our younger daughter and she has more of her father's adventurous spirit. This could be seen at the age of eleven when she was willing to do many of the things that Kelly and I shied away from. She could use the chain saw and haul wood. She chose to hike out to the island one of the first winters with her dad and her uncle. She is now a middle school guidance counselor with summers free to enjoy time and projects at the cabin.

While there have been many visitors to the cabin, there have been a few who have been regulars in helping to create this island getaway. Diana and LeRoy Anderson came up from the beginning; LeRoy to work on projects and Diana to help me "guard the lake" (we came up with that to justify sitting on the deck reading and talking.) When Diana died of lung cancer in 2003, LeRoy brought some of her ashes up and buried them under that deck where she loved to sit.

Brenda Fortier-Dube, a former student of Tom's, came from the beginning, hiking over in the winter, hauling logs during cabin building and then has come yearly ever since.

Holly and Lenny Charron became regulars when we moved next door to them in Hopkinton and discovered a shared passion for the out-of-doors. They fell in love with Toothaker on their first visit in 1997 and continue to come whenever they can.

When we first bought the land, one of our dreams was to have this be a place for children and grandchildren to visit. We now have Emma aged 7, Cooper aged 5, and Regan aged 3, spending weeks at a time here in the summer. All of them began coming when they were just a few months old; in fact Regan was born in late April and spent five weeks at the lake in August. They have added a new dimension to the cabin visits that Tom and I love.

Emma, Cooper and Regan

3 ~ A History of the Mooselookmeguntic Area

Being a history buff, I have spent many hours researching the Rangeley area and our lake especially. It gives me a sense of place and a connection to the past as well as being part of an ongoing community of people who have made this area their home.

The first settlers to the Rangeley region arrived in March 1817. Luther Hoar left Avon, Maine with his wife Eunice and 8 children. They were aiming for the perfect conditions, warm enough so they wouldn't freeze but still enough snow to pull the moose sleds the 26 miles to the rough cabin Luther and his sons had built during the previous summers. Luther had loaded these flat bottomed wooden sledges with all their belongs including the family bread bowl into which they had tied their infant daughter. At some point during the trip, the bowl, with the baby, fell off and put the family into a panic. They halted their trip while everyone began looking for her. She was finally spotted by one of her brothers, sound asleep in her makeshift bed.

Hoar was quickly followed by John Toothaker and his wife Deborah Raymond, who came up from Portland that same year. His wife was originally from New Orleans…one can only imagine what she must have thought of the long cold winters and the primitive accommodations.

It wasn't until 1825 that Squire Rangeley came to what was then known as Oquossoc Lake. He had been born in England in 1772 and came to the United States as a young man. He became a citizen in 1795 and in 1796 he inherited 30,720 acres of land that his father had purchased with three other men for $5546. Rangeley returned to England where he married and had six children. In 1821 on one of his many crossings to America (this was trip #7), he decided to move his family to Maine. He bought out his partners, brought his family over in 1823 and finally settled in Rangeley in 1825. When he arrived, he was greeted by Luther Hoar and William Toothaker. These two men along with others had built him a house. He was well loved by the town and, as the richest man, was given the title of Squire.

In 1827, the Rangeley's daughter came down with influenza and died. This was extremely hard on the family, especially his wife Mary. Not only was she living in a place where she had no educated companions, she also was in a place with no medical facilities. And so, while they stayed in Rangeley until 1840, Squire Rangeley gradually sold off his land until he only had seventeen hundred of the original thirty thousand acres left. He left Rangeley to move to Virginia where two of his sons had settled and there he stayed until he died in 1860.

Rangeley had sold much of his land to a man by the name of Daniel Burnham who, when Rangeley left, became the new squire. From all accounts, he was as different from Squire Rangeley as he could be. He was known to be thrifty, mean and ruthless. So even though Rangeley had left the town fifteen years earlier, the town fathers chose his name to give to the town upon incorporation.

The town originally was a farming community. But two other industries gradually gained in importance—logging and tourism. And this is where our own Lake Mooselookmeguntic begins to be important. Mooselook is located in both Franklin and Oxford Counties and is part of the Androscoggin River watershed. It receives water from several sources. The Cupsuptic River flows into Cupsuptic Lake, which is directly connected with the northern part of Mooselook. The Rangeley River and Kennebago River both flow into the northeastern part of the lake.

The lake's waters flow out to the southeast, into Upper Richardson Lake through "Upper Dam." The fifteen hundred foot dam was begun in 1845 and was in place by 1856 and was built expressly for sluicing logs. In 1885, it was rebuilt and made higher. It raised the level of Mooselookmeguntic Lake about fourteen feet, causing it to become joined with Cupsuptic Lake, forming a reservoir. One wonders what our island looked like before the dam was put in. One article I read speculated that one of the islands in the lake was probably twice the size it is now!

Upper Dam

Here is what Charles Farrar's 1876 <u>Guide to the Richardson and Rangeley Lakes</u> had to say about the dam.

> *The Upper Dam is one of the largest and most substantially built in the state; it is built of the heaviest kind of timber, bolted with iron, ballasted with immense rocks and is fifteen hundred feet long. In the spring when the water is held back for the purpose of running the logs down the Androscoggin, the pressure against it is immense. It is carefully watched both night and day, for should it ever happen to "go out"; the damage that would be done would be incalculable. The fate of Lewiston and Auburn and other places on the Androscoggin River, is really held in the timbers and stone of this dam.*

Mooselookmeguntic Lake's elevation is 1,467 feet above sea level with an area of about twenty-five square miles, making it one of the largest lakes in Maine. It is about twelve miles from Bemis to Cupsuptic

Campground. In the early years, it took about sixteen hours to haul logs from one end to the other. They used barges to boom the logs down the lake. I have a copy of a 1945 movie that was made by the Brown Paper Company of Berlin that shows the early days of logging. On it, there is a scene of a log boom being winched down Mooselook from Bemis to the dam. In it, you can see how these barges were used.

You can still see the skeleton of one of the barges, the HP Frost, beached on the spit of land on the south shore. It was bought and given to the Rangeley Historical Society. They, in turn, traded it for some antique lamps. The new owner moved it up by the causeway, which is where it was when many of us first bought our land. The owner was told he had to move it. His solution was to pull it down and beach it across from Toothaker on state land. Those of us on the south side of the island found it an interesting sight. But then one spring, we arrived to find it was down on Peter Axelson's spit of land. We were told that the state demanded it be moved so the owner just set it adrift. Different people have had ideas for it. Apparently at one point, someone thought to use it to haul cars out to the island. This person obviously envisioned a much more elaborate island community than what it has become!

While some logs were sent through the dam and down Richardson, other logs were sent to Bemis. Logs from Rangeley Lake were boomed across that lake, through a dam at its end and then down the Rangeley River into Mooselook. From here they were boomed down the lake to Bemis where they were loaded onto the railroad cars or went to the Bemis dowel factory.

Today Bemis is a boat landing. In the late 1800s and early 1900s, it was a thriving town. It had a dowel mill (which worked mainly with birch logs and employed 100-150 people), dance hall, boarding house, block mill, school, blacksmith, tourist camps, and a roundhouse and turntable for the railroad. It also had the only log railroad station in the world, located just beyond Bemis Landing.

In 1895, the railroad hired three hundred men to put in the railroad tracks from Houghton to Bemis. The railroad began in Rumford Falls and followed along the east side of the Swift River until it reached Houghton. Then it crossed the river to the west side where it followed the current paper company road to Bemis. In 1895 a station was built in Houghton and there was a station at Summit (the high point of land on the road) with a siding to hold twenty-seven cars waiting to make up log trains. Due to the steep climb up to the summit, two engines were often needed. Until 1903, Bemis was the end of the line but that year it was extended to Oquossoc. The railroad ran until the flood of 1936 washed out the bridge over the Androscoggin in Rumford. Because of this, the railroad had to look at the feasibility of continuing the line. With the increase in gasoline-powered trucks and the decline in tourism, it was decided to discontinue the line north of Rumford and the rails were removed in 1937. The WPA built the road from Houghton to Oquossoc in 1936-37.

While the railroad was initially established to transport logs, it also became an important part of the growing tourist industry. As hard as it for us to imagine today, people were coming to this area from Rhode Island and New York City to fish as far back as the mid-1800s. In the early days, the amenities were few. According to Edward Ellis, in 1864 the village of Rangeley consisted of two houses, three barns and a blacksmith's shop. By the late 1870s, the town had grown to twenty dwellings, two stores, a post office, two blacksmith shops, a carriage shop, a boat builder, a sawmill and one hotel! While the town was growing so were the sportsman's camps. The problem was getting to the camps. A small gauge railroad was put in that went as far as Sandy River, and from there travelers could take a mail coach or a stage. It wasn't until 1891 that the railroad reached Rangeley.

Some of the first "vacationers" to Mooselookmeguntic were a group of Yale students who arrived in the 1850s to spend the summer on what came to be known as Students Island. They built a cabin and spent the summer there.

The Oquossoc Angling Association was one of the earliest sporting camps. It was started in 1868 at Indian Rock, which is located where the Kennebago River flows into Mooselook, and is one of the few still in existence today. You can see it as you cross the Kennebago River on Rt. 16.

Vacations in the late 1800s meant packing up the family and arriving for a number of weeks at one of the many hotels or camps in the area. The Mooselookmeguntic House and Cabins at Haines Landing began in 1877, and The Upper Dam House was built in the 1880s and ran until 1958, when it was torn down. The cabins that remain were once part of this thriving sportsman's camp. Two camps that are still in existence are Lakewood Camps (1860) on Richardson Lake, and Bald Mountain Camps (1897).

While there were many camps in the Rangeley area, the most famous on our end of the lake were those run by Captain Fred Barker. Those folks lucky enough to own a copy of his 1903 book, <u>Lake and Forest as I Have Known Them,</u> get a fascinating look at what life on the lake was like in the late 1800s.

Barker was from Andover, ME and much to the horror of his family, he chose to make his living in the woods. He had tried living in Boston, but found he missed the Maine forests, so he returned and first appeared on Mooselook in 1870 while he was still in his teens. He was willing to do just about anything. He became a cook for a logging crew near Wilson's Mills. He would spend part of the winter ice fishing on Richardson and also worked as a river driver and a guide to fishermen. He was able to supplement his income by rowing people and supplies from Upper Dam to Indian Rock and back (15 miles) or to South Arm and back (24 miles).

In 1880, he bought a group of old fishing cabins at the mouth of Bemis Stream and tore them down and built Camp Bemis, which was known for the cleft rock in front of the dining hall. In 1885, he bought Students Island and built The Birches, which ran until much of it burned down in 1925. With the increased interest in the area, he decided to build The Barker in 1901. This was in the main part of the lake on Sandy Point near where Bald Mountain Camp is located.

In 1877, he purchased the first of seven steamers he was to operate on the lake and it was from running these steamboats that he got the title Captain. With the coming of the railroad to Bemis, he was able to meet passengers and take them to their destination on the lake or to meet other steamers operating on Rangeley Lake and Richardson. Jean Noyes, a long-time Rangeley resident, tells of stories she heard at the Rumford Bank when she first started there. In the early days, the tellers would get on the train for Bemis as soon as they finished work for the weekend. They'd arrive in Bemis, be met by Captain Barker and taken to The Birches on Students Island. They would then be picked up very early on Monday morning to get to the bank in time to open. That worked well except for the time the steamer forgot them and the bank president had to run the whole bank himself until they finally arrived.

By the time Captain Barker died in 1937, the heyday of the large resort hotels was fading. There were a number of reasons for this. The railroads had discontinued service and the tracks had been torn up. WWII focused people towards the war effort and when the war ended, the public became more mobile and liked to travel around more than stay for extended periods in one place. Or else they wanted second homes so many of the camps were split up and sold to individual families.

This leads me to Shelton Noyes and his role in the development of the lake. In 2007, I sat down with his wife Jean who shared memories of her deceased husband. He was a lawyer, a former state representative, a banker and a realtor. According to her, Shelton grew up in Rumford and she laughingly told me how people in that city thought he was crazy because of his wild ideas. In 1955, Shelton was president of the Rangeley Trust Company and was receiving frequent inquiries from people who wanted to buy land in the area. Most of the land at this time was held in huge blocks and it was hard to buy single lots. So Shelton bought most of Rangeley Plantation, which turned out to be thirty-six square miles. His vision was to offer lots to people who wouldn't ordinarily be able to afford them. He started by building cabins and selling them, but found that what he'd built and what people wanted were never quite the same and so he ended up just selling the land instead. In a brochure he had printed at the time, he was offering land for ten dollars down and ten dollars a month.

And now we come to the history of our island. The long and short of it is there is just not much information available! Other than it being noted as having huge trees or people passing by on boating trips, Toothaker Island is rarely mentioned in any of the history books. The island was included in Shelton's land purchase of 1955 and he is the one that many of us on the island were fortunate to get to know when we bought our own pieces of land. Even 30 years later, he was still eager to do whatever he could to help people realize their dream of a vacation home.

In the last twenty-five years, Toothaker has seen changes as it has been divided and subdivided and built upon. But it still retains its character and the people who now own it all have an investment in keeping it a place to find refuge from the often-hectic lives we live in our home communities.

And so life continues. What happens tomorrow is the future but what happened yesterday is already part of the on-going history of the island.

4 ~ A Tent is Not Enough

BUILDING A CABIN ON AN uninhabited—some would say uninhabitable—island on a lake offered Tom and me challenges that we never could have envisioned. What started out as an easily imagined adventure turned into problems and frustrations mixed with laughter and fun. Finally, through the efforts of many good friends and family, we ended up with a beautiful log cabin and a deeply satisfying sense of accomplishment.

When we first bought our fifty acres of wilderness, there was nothing there but an overgrown and tangled spruce forest. We spent the first year clearing a space for a tent platform and some basic living space. We had twice camped our way across the country so we had lots of gear that could now be used at our campsite. We put our tent up on the tent platform, Tom built a picnic table that we hauled across in pieces and then assembled and we created a fire pit to cook our meals. We also had our old Coleman stove and lantern. We nailed a small wooden cabinet between two trees and once again made use of the "camping box" that Tom's mother had made back in 1955 when she drove her four kids from NJ to CA. We were all ecstatic when we were able to get a huge wooden box over to the island because then we could leave things from one trip to the next.

A lot of this early time was spent doing two things. One was clearing all of the brush and blow-downs and taking down trees at the cabin site. The other chore was sitting on the beach, picking up rocks and tossing them into holes where the water was eroding away the underpinnings of our campsite. In those early years, I even thought I could clear a beach area that would be rock free. I finally gave up on that idea when I'd return every spring to a new crop of stones littering the sand. We did have a few friends (usually male) who loved the idea of moving BIG boulders, and they would spend whole afternoons on the beach with levers, shovels, pulleys and chains.

As we continued to clear, we were also researching suitable cabins. We looked at shelter kits that were prefab boxes and visited or sent for information from many log cabin companies. We finally settled on the log cabin we wanted and ordered it. One of the things that attracted us to this particular company was the fact that they offered a log cabin building school. Tom and I planned to go in May and then build during the summer. However, by the time May rolled around, I had switched jobs and would be working all summer. Tom Lajoie, a former student of Tom's, had volunteered to help build the cabin, so it seemed sensible for him to go to the weekend training in my place. They returned disappointed in what they had learned but still eager for July 15 when the first load of logs would arrive. They were confident that they could make up with common sense what they lacked in skill.

The first job that spring was to put in footings…the ten cement posts that would hold the cabin. We arrived at the island feeling smug and professional with our post-hole digger, line level, cardboard sauna tubes and bags and bags of Sakrete.

We dug our holes, no easy job in the rock-infested uneven ground that was our building site. The post-hole digger itself was enough to discourage anyone. With its giant blade, long arm and engine, it resembled a leftover kitchen appliance from the days when Paul Bunyan roamed the Maine forests. Using it was like being on someone's sick idea of an amusement park ride. It would shake you up and down until it hit a rock at which point it would stop…and you wouldn't.

When the holes were dug, we put in the forms and began to mix and pour…and mix and pour…and mix and pour. There had appeared to be an endless supply of seventy-five pound bags of cement when we were carrying them up from the dock. Now with less than half of our forms filled, we were faced with a very finite number. We realized that our careful calculations for how much we'd need were way off and an emergency trip was launched for more.

As each post was filled, we carefully leveled it with the one before until finally the last one was finished. We stretched the line level from the last post back to the first post and were dismayed to find a seven-inch difference. But by that time we were once again out of cement, out of time and definitely out of patience.

The following weekend we returned with more cement and an even more professional piece of equipment. From somewhere we had come up with a surveyor's transit that would allow us to make more accurate measurements. We decided that the first post was our high mark and began to add cement to the other nine posts to make them level with the first. This meant just adding an inch or two to some posts. On others, we needed to tape on extra sauna tubing to gain the height we needed. When we finished, we sat back with a sense of satisfaction of a job well done. But ever cautious, I suggested we check our measurements one more time. We were dumbfounded at what we found. Now, instead of being seven inches between the first and last post, there were fourteen! We had, in our enthusiasm and ignorance, done everything backwards. The last post was the highest––not the first. We quickly ran from post to post, pulling out the wet cement with spoons and hands, continually adding water to keep it from hardening. And then we began all over again. This time we got it right and when July 15th came, we finally had a level base on which to build our cabin.

Unfortunately, during the week before the initial delivery, Tom threw out his back and became totally bedridden. We panicked as we realized that in seven days he was going to have to start hauling 35,000 pounds of logs and boards out to the island. He tried ice, he tried heat. He took drugs and got massaged. He went to a physical therapist. We have never been sure which of these things worked but by the 15th he was back on his feet and ready to go.

Tom Lajoie and the logs arriving from Josselyns

Probably the worst part of building the cabin was hauling materials over two miles of unpredictable water. The few times I had to do it made me thankful that I was working that summer and unavailable for daily duty. The best time to haul was from 5 a.m. to 9 when the wind usually came up. Each trip involved loading about ten logs onto a raft made of four canoes lashed together, and while one person drove, the other one watched and guided this unwieldy makeshift barge. Once at the island, it needed to be landed in a way that made the logs accessible. Then began the process of carrying the logs up to the building site. The cabin took about two hundred fifty logs, so the hauling was backbreaking and endless.

Canoe raft used to haul logs.

While the mornings were spent hauling logs, the afternoons were spent building the actual cabin. My job was to come up for the weekends to resupply the kitchen and to carry a few…a very few…logs. Each time I arrived, the walls were higher and it looked more like a real cabin.

Various people helped with the building. The most important was Tom Lajoie, a former student turned friend. With his long blond hair, laughing eyes and red bandana, he looked like a leftover from the late sixties. But at nineteen he had more carpentry knowledge than Tom had managed to acquire in forty-three years. We were touched and grateful at his willingness to give us three weeks of his summer. "Old Tom" and "Young Tom" were a good pair and together they built and cursed their way to a finished cabin.

Brenda appeared for a week and got involved in hauling logs. She still feels that she risked her life jumping off the moving boat to try to help Young Tom save the logs that were floating off the sinking canoes. Over time, they perfected a method of landing the raft so, remarkably, it ended up right where they wanted it on the shore.

At the end of three weeks the cabin was up to the second floor, and Young Tom had to leave. Tom continued working on his own until he got to the roof rafters. When measuring for these, he realized that the cabin was out of square by four inches, which may not seem like much, but he guaranteed me that it made making the back roof rafters impossible. So we called on Keith King. Keith is a person who does not acknowledge the word impossible. And sure enough, he showed up and quickly figured out what to do.

Tom's brothers and their families came in late August, and that was the weekend we put on the first layer of the roof. (Don't ask me why but this roof had more layers than I thought possible.)

Over Labor Day our longtime friends and NH neighbors, LeRoy and Diana came up with us, and we continued with more roof layers. It was fun to watch our daughter Lindsey and her friend Sarah pounding nail after nail while Leroy supervised us all, moving his six foot-four-inch frame as effortlessly over the steep pitched roof as he did on the ground.

Cabin up to the rafters

Cabin at the end of the first summer

By the time we closed up for the winter, the windows were in and Tom felt he had met his goal of ending the season with a cabin that was enclosed. We were lucky to have had so many willing hands. The cabin stands today as a monument to creative ingenuity and hard work.

5 ~ A Simple Cabin is Not Enough

We started building the cabin in 1987 and finished it over the next couple of years. Its dimensions were 24'x24' on the first floor, and 24'x16' on the second story. The downstairs was open except for a small corner bedroom; the upstairs had two bedrooms. Once it was done, we could breathe a sigh of relief. We had a place to get out of the rain, get warm and relax. But it wasn't long before we started thinking about changes and additions. It seemed like we never had enough room for people or supplies, so someone was always making observations or offering suggestions that would perhaps remedy these problems. We never knew when one of these conversations would lead somewhere.

One of the first things we added was a bunkhouse behind the cabin. It started out as an unfinished room with a double bed and two bunks, but then we decided to finish it off with insulation, pine paneling and a reorganization of the beds. When we finished, we all decided it looked more like a motel room than something you'd find behind a rustic cabin. But anyone who has stayed in it thinks it's terrific.

One of our major inside projects began one Fourth of July. Our friends, Barb Steele and Ben Skaught were visiting from Connecticut along with Linda Sweeny who had come up from Massachusetts. Our project for the afternoon was to put in a piece of Formica countertop. I had mentioned to Tom how nice it would be to have a counter between the stove and the end of the regular counter to give us more workspace. As Tom and I were measuring, Linda walked through and asked an innocent but fateful question. "Could you move the work counter back a few inches toward the bedroom? Then you would have a little more room in the kitchen." I don't think she expected to be taken so seriously, but her words got me thinking of how crowded our kitchen was whenever more than one person was trying to work. Something--or someone--was always being bumped, stepped over or knocked into.

"Hmmm," I thought. "What would happen if…?"

Tom, seeing the look in my eye, tried to ignore me as I began to talk. To him, the small kitchen wasn't a problem. He usually cooked alone and found the easy reach of things perfect. But, he agreed, it did get cramped when you added another person or two. So we began to measure and look at what kind of difference a few inches would make.

Linda, at this point, disappeared while Barb and Ben, seeing the fun we were having, decided to get into the act. Being flatlanders from the city, they think big and act fast. Before we knew what hit us, they had a plan that meant moving the bedroom door, swinging the work counter around against the back wall and moving the upper cabinets above it. With a little more careful measuring, they had space for the stove along that wall as well.

Tom tried making a weak protest having to do with his tools and where he would put them--for you see, the work counter was actually on the backside of the refrigerator and was his designated "shop" area. But his protests were brushed aside as the rest of us got caught up in the grandiose nature of the plan (except for Linda who was now asleep in the hammock, unable to watch where her comment had led us).

We finally were able to assure Tom that he would have more and better space for his tools, and that while we had all these willing hands, we had better get started.

And so we did. By the end of the first hour, the door to the bedroom was moved to the opposite end of the wall. Then we slid the work counter over and unloaded Tom's tools while continuing to offer him ideas as to where he could put them.

Next came the unloading of the upper cupboard. We unearthed various sticky treasures and some very old food. Barb kept a pot of water heating and scrubbed dishes, shelves and boxes as we went. When the cupboard was empty, Ben began to unscrew it from the post. He left two screws in while he went to take some screws out of the wall end.

The cabinet decided that three screws were not enough and fell with a crash onto the refrigerator and Ben. We quickly hoisted it up and swung it around to the back wall where, to be safe, we put in twenty screws.

After refilling the cupboards, I stood back to look at our new kitchen. It seemed to have some real advantages, but I was feeling a little guilty over Tom's tools. His promised space did not seem to be materializing in quite the way we had hoped.

The next morning, Tom and Ben moved the stove. I loved the way it fit right in between the two counters and opened up the rest of the kitchen.

There was only one job left: putting up the pegboard so Tom could hang his tools. Linda suggested he build an outhouse, but quickly realized that "outhouse" was probably not a good choice of words. We talked about using hinged boards that could be pulled up, little carts that could be moved around, but finally settled on putting a small piece of pegboard above the old work counter, between it and the upper cupboard, and then a larger piece behind the refrigerator. The final result was that the tools were much more accessible and we no longer had to look at the ugly back of the refrigerator.

While I pondered other inside changes, Tom continued to mull over where he might build a real shop. He decided that what he needed was another building, and so he began to wander around the back of the cabin muttering to himself that maybe this would be a good spot or maybe that area would have just the right rocks for posts. Finally he settled on a spot to the right of the path that he thought would be perfect. Well if not perfect, at least possible, with places for footings and no large trees to cut down or fall down.

The following summer, his goal was to get the deck finished. He claimed it was because the building permit was going to run out, but I think that was just a ruse to get this project underway. By the end of the summer, a 16' x 20' deck was down and painted and ready for winter.

During the next winter, Tom began to seriously plan for his long awaited workshop. Lenny took us to a discount building supply store where we bought a door and six windows of varying sizes. Tom Lajoie, our carpenter extraordinaire, went over the plans and together they figured out what wood was needed. This was all delivered and hauled across the lake by Lenny and Tom the first part of July.

Finally on August 1, it was time to start the actual building of "THE SHOP." Tom had his tool belt ready, he had Lenny with his tape measure and skill saw….and most important of all, he had Tom Lajoie with his knowledge and nail guns. By 8:30 a.m., they were set to go.

I was unprepared for how quickly this shop would appear. By 11:00, the first wall was up, by 12:30, the second wall was standing. By mid-afternoon, they had all four walls up and Lenny was cutting the rafters! By suppertime the rafters were up and the roof was laid.

Problems were few. One window got framed in sideways and had to be redone, and Tom and I had to run to the neighbors to "borrow" four 2x6x18' boards as we had run short.

Holly and I did not sit idle. Besides our usual daily job of guarding the lake, we kept the three workers fed. Whenever they seemed to be flagging a bit, we hurried out with some Twixt bars or cheese and crackers.

The second day started out about 8:15. There were only two things left to do…shingle the roof and put in the windows. Tom and Lenny lugged shingles and tossed the bundles up to Tom Lajoie, who threw them onto the roof and nailed them down. The whole process seemed to take about thirty minutes and the roof was done. Then came the windows. The only tricky part was that because they were varying sizes, each one had to be carefully measured so it would match with the correct hole. When they got to the window that had been reframed yesterday, they realized that the original hole had been correct. So we now have a window that opens horizontally that should open vertically.

The door was the last thing to go in and the building was done. It was 11:30 a.m. Tom L. left and Tom and Lenny rested. Holly and I continued to guard the lake.

Once the shop was done, it was time to start planning what would come next. Tom and I decided that we needed a bedroom on the main floor. The upstairs was stifling in the summer, and so by putting in a downstairs bedroom, we would solve that problem as well as create more space for guests. It would also prepare the cabin for our "senior years," when getting up the stairs would be more difficult.

Tales from Toothaker

The shop at the end of the weekend.

    Both the bunkhouse and the shop are traditional stick built structures. But an addition on the cabin meant MORE LOGS. This time, however, we had a new boat launch to haul from, which was a mere half-mile away. Kelly and Todd were ready for hauling duty. Kelly made sure we got lots of pictures of her doing AML…awkward manual labor. Hauling those logs off the raft and up the hill to stack them was not only hard but awkward as well.

    Tom had learned a lot since originally building the cabin. For example, rather than pour cement posts to rest the foundation on, he and Lenny poured two inch high concrete slabs and then bought 6"x6" posts to rest on top of these footings. However they faced one new problem. They needed to find a way to attach the logs into the existing cabin. Between the two of them, they came up with a plan--and it seemed before we knew it, we were moving into a completed room.

    You might think at this point we'd be done. We had the bunkhouse for guests as well as two upstairs bedrooms. The kitchen was well functioning and Tom had his shop. However, there was still a problem… noise. Tom loves quiet afternoon naps and both of us go to bed fairly early. But our one small gathering space was right next to our bedroom— and even the upstairs rooms which now housed sleeping grandchildren heard everything going on below.

So we began to think about adding on a family room. After much discussion, we finally settled on the perfect spot. Tom's shop! Yes, once again, he and his tools were being uprooted. But after an initial protest, he began to see the advantages. He would get a new improved working space that would have more light, more useable space and a level floor. Plans were made for the new shop off the back of the old one and the following spring, it was constructed. He put in a huge window, which let in plenty of light. He affixed pegboard everywhere to hang his tools and stuck rakes and shovels and hoes up in the rafters. He built a big shelf over the window where he stores the kerosene heaters and extra insulation. There are narrow shelves along one wall that hold every conceivable sized nail and screw. I dare say he thinks it is the best one yet and is glad that he is finally settled. Of course he does not hear Kelly and Lindsey whispering how this building would make a great second cabin.

Once the new shop was done, the next project was a screened porch off the lakeside of what had been the shop and what was to become the family room. While bugs are not a huge problem due, we think, to how windy it is, they can still be bothersome. Plus it would be a nice spot on a rainy day.

This turned out to be one of the easier projects. Tom and Lenny put on a deck, some supports and a roof and then screened in the whole thing. Our friends down the lake gave us a set of French doors to use to enter it from the family room. Keeping in line with our philosophy that most of our stuff is at its final stop before the dump, we outfitted the porch with hand-me-down furniture and then found it a perfect place for an afternoon nap, for an extra visitor to sleep or a play space for the grandchildren.

While this was going on, the family room remained a dark, unattractive space that was used to store recycling, garbage and miscellaneous swim toys. Finally it was time to turn this uninviting room into something people would use and enjoy. It started with adding two skylights for additional light. Immediately the room had a whole new feel. The roof was insulated and the ceiling covered over with thin plywood. The walls were next. These we insulated as well and then covered with tongue and grove pine paneling. While Tom was at it, he did the same thing to the inside wall of the porch. We hooked up a wood stove, the kind where you can see the fire if you wish, added some gas lights, put down a rug and gathered up more used furniture. The final touch was hanging a beautiful silkscreen painting that our friend Karen Kitzmiller had done before she died. Her husband Warren and his new wife Jeanne hiked down from their cabin to present it to us.

The room is comfortable with a very different feel from the cabin. And we all agreed it paid for itself the first weekend it was finished when it housed five active kids during a prolonged rainstorm.

Tom and Lenny listening to the ball game in the new family room.

The most recent addition is for Lindsey. Since Kelly and Todd and the kids have taken over the upstairs, and other guests are often at the bunkhouse, Lindsey tends to get shuffled around to wherever there is a bed. We decided that she needed her own space. We first found a place for a small cabin back in the woods. Tom cleared the space and made a trail that went from behind the outhouse to the new site, which also connected to the shore trail. However, Lindsey felt it was too far away, and Tom decided that hauling the wood to the site would be difficult, so instead he settled on adding a room onto the bunkhouse. We got windows and a door given to us from someone who was redoing their family room and two more windows and some inside siding from our friend Peter Axelson. And the trail that Tom made to the originally proposed site is in constant use by the grandkids.

Bunkhouse and Lindsey's addition

I haven't dared ask what will come next. I can only wait and see what new plan emerges to keep Tom challenged and busy.

6 ~ Outfitting Our *Simple* Cabin

Part of the challenge of a cabin such as ours is finding ways to do things that we don't even think about at home. Everything is more of a production at the lake…from furnishing the cabin to figuring out how to wash the dishes. Some people would be discouraged by all this work. But we see it as part of the fun.

Furnishing the cabin is a process that has gone on for years. Here again, friends have helped us out. One couple was redoing their kitchen and gave us their old cabinets. Someone else gave us an old table and chairs and passed on some old braided rugs. We got a small gas stove and an old gas refrigerator through friends where Tom worked. Yard sales produced more treasures. We got our stainless steel kitchen sink at one in New Hampshire and, while on a visit to Minneapolis, we found eight purple plastic plates with pictures of Vikings on them as well as two old gas lanterns that a man was selling as ornaments. He obviously did not know that you could send them back to Coleman to be reconditioned for a very reasonable fee.

Finding the things we needed was only part of the problem. Getting it over to the island was a whole different challenge. Everything had to come over either in the fishing boat or in our eighteen foot Starcraft. Some things were just awkward…like the three-piece sectional sofa that we had inherited from Tom's dad. Others were not only awkward but also brutally heavy. Everyone who lifted and carried the original refrigerator will attest to that.

Moving the refrigerator through the sand at Indian Cove.

Not having electricity meant that we have to rely heavily on propane gas, which we buy in 100-pound tanks that we pick up in Rangeley, load into the boat and then haul over. Propane runs our stove, our refrigerator and the installed gaslights.

Our first refrigerator was a Servel that was found in someone's back yard. These refrigerators were sold from 1926-1956 and although marketed to the general public, they were used mostly by people who did not yet have electric power. While our Servel kept things cold, it would not keep ice cream frozen. And although it was big on the outside, the actual storage space was small. Gas refrigerators have since been updated due to their use in motor homes and we considered replacing ours with one of these, but decided to wait. Then my mom came for a visit and decided that we should buy a new more efficient model and offered to pay half. So now we have a much smaller unit that holds a lot more and freezes ice cream!

We also use gas for lights. We have the old Coleman quick lights that were used in homes in the early 1900s and run on white gas. These are even more fun to start than the gas oven. You pump them up, turn on the gas and strike the match. As soon as the gas ignites, you turn it off and wait while flames shoot out of the top and the gas turns into a vapor. At this point, you turn the gas back on and watch the mantles glow. A few more pumps and there is enough light to read by.

A number of years ago, we decided to buy five gaslights that we could run off of the propane. Tom and Lenny ran gas lines to each site and attached the lights. These are much easier to operate. We turn on the gas, light a match and the mantle is instantly bright. However as we get older and the 100 pound tanks feel heavier, we are back to using the quick-lights as a gallon of white gas is much easier to bring over. We do have a generator for those times we need electricity which usually involve power tools, although I must admit I use it to run the vacuum cleaner and to pump up our airbed.

Decorating the cabin has been fun, as most of the things we put on the walls have stories that go along with them. We have a hooked rug that Tom's grandmother made that says, "Highland Farm," which is the name of the family get-away in New Jersey. A few years ago, we got another piece of farm memorabilia when Tom's niece sent us a large carved wooden sign that says, "Escape with the Herberts." "Escape" was the name of his grandmother's house at Highland Farm. We also have a painting done by Tom's stepmother that is of the pine grove at the farm where his mom and sister were originally buried.

We have found many interesting items in the bottom of the lake that we have rescued and mounted. As mentioned earlier, Mooselook was a log highway. For some reason, our particular piece of shoreline seemed to collect a lot of debris from when the logs were boomed down the lake. We have a cant hook (a wooden handled tool with a moveable metal hook on one end and used to move logs) on display, as well as a metal spike. But by far the most plentiful items were boom chains: five-foot pieces of heavy chain with a large ring on one end and a flat bar on the other. These were dropped through holes in the boom sticks (long logs), thereby allowing the sticks to be attached together. We have given these chains as gifts, donated one to a logging museum in Northern Minnesota and have one hanging on our wall. I had one in my classroom to use when we talked about early logging in New Hampshire. There is even a lodge on Moosehead that uses them to hang beds from the ceiling.

We have found a lot of interesting driftwood. On one piece rests the ball of the first dog we had at the lake. She loved chasing it, and when she died, Tom decided to put it on a driftwood "sculpture" that we had hanging on the wall. We have an Old Town canoe paddle that was presented to us after we rescued some overturned canoeists. There is an elevation map that is three-dimensional, so it shows all of the surrounding mountains. In another spot is a clock made by Tom's oldest brother many years ago. He used an interesting slice of a log for the face. A deflated red Mylar balloon has recently been added after being discovered one winter deep in the woods.

One Christmas Tom made me his version of a printer's box. In it we put small "treasures" that we find: unusual stones, feathers, skeletons. Anything that catches someone's eyes.

Wall decorations.

Besides decorating, we have added conveniences that never would have been available to early pioneers.

One of the things that bothered me the most in our early days was not having access to a phone. At home, I hate the phone, and will go to elaborate lengths to avoid answering it. And I didn't want a phone up here for anything other than an emergency. So, as soon as we had access to a cell tower, we got a cell phone. I feel better knowing that if our family needs to reach us, they can try—and with good weather and a strong signal, they might actually get to speak with us. But, if not, at least they can leave a message.

Water is one of the easiest as well as one of the hardest things to get. While there is plenty available, bringing it up in buckets takes strong arms and a firm step. We use lake water for dishes and cooking. We originally drank from the lake, but then went to hauling our drinking water from the mainland. We used two five-gallon containers that we'd fill up at home. If we ran out, we could drive in to Oquossoc, where there is a natural roadside spring. Then one day, while looking through a Lehman's catalog (known for non-electric tools and appliances), we saw an ad for a British Berkefeld water filter. After learning how well it purified the water, we decided that it was a must-have addition to the cabin. We can use lake water that we filter through the "Big Berkey" and have a steady supply of clean drinking water.

Bathing in the summer is no problem. We jump in the lake whenever we need to clean up. Winter trips are tougher. Sometimes we just do the best we can with a washcloth at the sink. But our favorite method is an outdoor shower on a sunny day. One of us stands in a dishpan full of warm water while the other holds and guides the sun shower. The sunshower is a black bag with a hose and showerhead attached. We fill it with warm water that is released as needed. Needless to say, these showers are fast.

Washing clothes is a bit more work than washing ourselves. While trips back to New Hampshire usually involve loads of dirty clothes, sometimes we run out of clothes between times. Kelly, with the three kids, seems to be in this position the most. We use a big plastic bin filled with lake water, and then throw in some biodegradable soap and the clothes. The easiest way to get them clean is just to jump into the bin and stomp around for a while. The clothes are rinsed in the lake and then rung out. This used to be the toughest part of the job. I was complaining about this to my friend Beth Ann Paul on her first visit here. She went home, and found us an antique hand wringer on EBay that is much easier to use and gets a lot more water out of the clothes. Tom has put up a long clothesline on a pulley that goes from the deck to a large tree, so there is plenty of space to hang out our wet clothing.

An outhouse up the hill meets our bathroom needs. Getting an actual building was a big step forward, as we started with just a box over a hole that was crude, practical and quickly built. It was also not that comfortable in the rain or snow, and lacked privacy. Thus, the metamorphosis to the outhouse.

Our outhouse is perched on a knoll about two hundred feet back from the cabin. It has windows on each side as well as a screen door. In the summer it has flowers or freshly cut pine boughs. It is also decorated with pictures and signs, one of which cautions users to be careful with the lime that we use to speed up the decomposition. Once it was decorated with a collection of stuffed pigs. It has had posters and poems. We are always up for a new look.

In winter we cover the screen door with plywood. This stops the snow from blowing in, although it does not help with the temperature. To make sitting a bit less painful, Tom cut a piece of dense foam pad in the shape of a toilet seat. This sits behind the woodstove getting nice and warm. When nature calls, the pad is taken and put inside one's coat. Upon arrival, it is plopped down on the ice-cold seat and comfort is achieved.

One of the problems with an outhouse is odor. The way we have dealt with it is to help the waste break down by moving it forward in the hole underneath the floor of the outhouse. The hole was made purposely large with a removable piece of flooring for easier access. But eventually it does get full. What to do next resulted in a difference of opinion.

Tom wanted to move the outhouse and set about digging a new hole. It was going to be the same size as the first one, four by four by eight feet. In the fall, he laid it out and began to dig. This is when the conflict started. I saw where he was putting it and realized that it was facing the wrong direction. Instead of looking at the lake, you now looked at the back of the cabin. And it was also in direct view of the bunkhouse. Winter came and we continued the discussion. When I had just about given up, Tom and Lenny announced that the outhouse would not be moved after all. Upon reflection, they realized that it would be too much work to lift and carry the building to the new site. Instead, they were going to dig out the old hole and use the new one as a receiving depot. This involved the use of shovels, a wheelbarrow, a rake and a hoe. However the best tool was the posthole digger. Eight wheelbarrow loads of basically odorless manure were moved from one hole to the other.

We have a book of outhouse recollections that often feature outhouses tipping over. My dad told a childhood outhouse story involving him, Halloween and some pranksters, in which he ended up in the hole when trying to go after the culprits who had knocked it over. Our outhouse at the lake has tipped over, but the culprit was one of the worst storms ever to be seen in the area. Tom and I were not even there, but Lenny and Holly had gone up for the 4th of July. The storm came through in the early morning and snapped off numerous trees, one of which landed on the outhouse, knocking it right onto its side. One of the windows was smashed but otherwise it was not harmed. Lenny and Holly's son arrived later in the day, and the two of them managed to right it.

We have a composting toilet that we keep trying to figure out where to put. On those cold dark nights in the winter, it would be nice to stay inside. But we can never seem to find just the right place to locate it, so it sits in its box under the cabin waiting for age or infirmity to push us into using it.

Most people who continue to return to the lake, find living with fewer amenities makes life simpler. It certainly makes us think about all of the things we take for granted…and how many things we could do without.

7 ~ Cooking on Toothaker

One of the most fun and satisfying activities on the island is cooking. Because there is no option of running to the store for something forgotten, the cooks end up being more creative than we ever are at home. And those eating are never sure what might appear on their plates…especially at the end of the season when we are trying to use up things that won't keep over the winter. But no one ever complains, and some of our food inventions have become legendary.

When we first came to the island, we did all of our cooking over an open fire or a Coleman stove. Campfire cooking has a certain romantic appeal, but is actually very time-consuming, and I never seemed to learn how to regulate the heat so that the coals lasted until the chicken was cooked all the way through. Our favorite campfire dessert is putting applesauce into the bottom of a cast iron kettle and then pouring a prepared gingerbread mix over the top of it. You cover the kettle and set it on the coals. The applesauce keeps the gingerbread from burning and tastes wonderful served over the top.

Once the cabin was built, we were delighted to have a friend give us an old apartment-size gas stove, and we most often use that for cooking. And even though it is probably 75 years old, in some ways I like it better than our more expensive stove at home. It is sturdy, heats evenly and can be adjusted to a very low flame. Its downside is that the oven does not have a pilot light, and so it has to be lit using a match, which can sometimes cause some rather dramatic loud noises and scare the unprepared.

We start the day with breakfast, which Tom cooks. His meals are always memorable and part of the lake mystique. He gets up, has his morning dip and then grinds the coffee by hand. Some mornings he makes oatmeal using regular oats (never instant), to which he adds a chopped apple, some walnuts, a handful of raisins and any other dried fruit that happens to be in the cupboard. He serves it with milk and brown sugar, and even die-hard oatmeal haters seem to like this.

Other mornings he makes pancakes. A few years ago we discovered that Arrowhead Mills makes an organic multi-grain pancake mix that only needs water added, so this is what Tom uses. The pancakes have a delicious flavor on their own, but become even better when topped with cut up fresh fruit and real maple syrup. Tom also puts peanut butter on his…a custom that cause many to roll their eyes in dismay. He claims that when he eats pancakes for breakfast, he is starving by ten. The extra protein in the peanut butter gets him through to lunch.

After breakfast, we usually sit around the table and finish up the coffee. The sun beats in the sliding glass doors and the waves lap the shore as we talk about anything from the mundane to the philosophical. Tom is usually the first one to jump up, anxious to get started on the day's projects, and so breakfast comes to an end.

Lunch tends to be the most casual meal of the day. Whoever gets hungry first goes in and pulls out cold cuts, cheese, veggies, pickles and whatever kinds of bread are around. Everyone makes his or her own sandwich, grabs a handful of chips and some juice and heads out to the picnic table.

Four o'clock is generally snack time. One of everyone's favorites involves using one of the most disgusting things in our kitchen, an old red three-quart saucepan. At some point in its existence it had a cheap coating of black Teflon, but that had worn away even before I got it. The top is still red on the outside and greasy brown on the inside. The pan itself is now mostly black with the occasional red streak showing through. The handle is black plastic with some electrical tape around it and a screw and grease-encrusted hose clamp holding it onto the body. And yet if we could only save one thing from our kitchen, it would be this. For this is the popcorn pan…and everyone loves it for the high quality popcorn it produces.

Tales from Toothaker

People do not forget island popcorn. I make it the same way I do at home (where I also have a popcorn pan, but not as good). I first put in some safflower oil and three kernels of popcorn. When these pop, I cover the bottom with kernels. I gently begin shaking while listening to the popping sound coming from within. Being the lid doesn't quite fit, there is always a small gap that lets the steam escape. When the popping slows down, I take it off the burner, take off the lid and dump the finished product in a big plastic bowl. I add some salt and take it out to whoever is here.

The response is always the same. "How do you make this popcorn? It is so fluffy, so tender." I explain how I do it, but always end up telling them that I think the secret is in the pan. But then our friend Caleb Morrill-Winter reminded me of the year I had taken the special red pan home for the winter and found that it just didn't work as well in New Hampshire as it did on Toothaker. So while I think the pan is part of it, it is not the complete explanation. But I am not really concerned as long as the popcorn continues to come out so well.

In the past few years I have added an element to the popcorn making that pleases Tom, who is the only person I know who does not like popcorn. I often will throw a couple tablespoons of sugar onto the kernels after I add them to the pot. This allows them to pop even larger as the sugar melts to a higher temp, and also coats the popcorn with a sweet saltiness that most people love.

By evening, everyone is starving again, especially on those days when we have been outside for long stretches. It is the one meal I assign to guests; I tell them to bring one dinner to serve x number of people. No one worries if we have the same things; it is a given that the same thing is never really the same. Like the weekend we had chili twice. Both kinds were vegetarian, but one had walnuts in it and the other had corn.

Everyone at the lake pitches in to help out in the kitchen. It is a rule that if you cook, you do not clean up. But we are never at a loss for a clean-up crew. Our friend LeRoy, in fact, would feel incomplete if he did not have a chance to do dishes during his visit. He is the only person we know who sees washing dishes as a sacrament and he does it with a reverence and joy that no one else seems to have for the job.

Desserts are our downfall! Where none of us eats desserts regularly at home, every meal at the lake ends in some kind of sweet treat. Holly is known for whipping up a pie, my sister-in-law Terry Monette makes "Best Cake", and Kelly and Lindsey will make anything that involves chocolate chips.

Our desperation dessert was created one night when we realized that we didn't have anything prepared. We began to search the cupboards and found graham crackers (always on hand for s'mores), some chocolate frosting in a can, and in the refrigerator, whipped cream. We smeared frosting on the crackers and piled whipped cream over the top and ate one after another. Tom, as usual, added peanut butter to his.

Another good example of kitchen creativity was the salad that Holly made on one winter trip. All we had was lettuce…and lots of it. There were no other vegetables and no salad dressing. But there were apples, grapes and leftover candied pecans that she mixed with the lettuce and then made an olive oil/lemon dressing.

While I always have a garden back home, growing food on the island has not been as successful. The season is short and cold; so many things do not do well. The most successful crop will come as no surprise to those familiar with Maine…potatoes. The year before she died, Karen Kitzmiller created some raised beds near their cabin and planted perennial flowers in one and potatoes in the other. She inspired me, and the next summer we had a bumper crop of white potatoes. There is nothing quite like digging up new potatoes, boiling them and serving them with melted butter and parsley.

Going out to eat while on the island is another adventure. It always involves either driving our boat to Bald Mountain Camp in the upper part of the lake, or going over to the mainland, getting in our car and then driving into Oquossoc or Rangeley. Our best trip, however, was the time our island neighbors John and Priscilla McAdams invited us to go with them to Lakewood Camps on Richardson Lake. Richardson is the next lake, and we can go to the dam on Mooselook, dock our boat and walk a quarter mile to Upper Richardson. So that is what we did the evening of our dinner. Waiting in his boat for us on Richardson was the owner of Lakewood Camps. We piled aboard and headed down into Lower Richardson, where we enjoyed a phenomenal supper.

Tales from Toothaker

Family dinner at Bald Mountain Camps.

After we ate, he took us back to the landing and we hiked back to Mooselookmeguntic and boated home.

There is a lot of sharing of meals amongst people on the island. There are annual island-wide gatherings on the 4th of July and on Labor Day. Most often, however, we have smaller potluck dinners. On one of our winter trips, Warren Kitzmiller and his daughter Carrie were going to be at their place, and we decided to get together for one of our suppers. I thought it would be nice to invite Bert and Chris, who live down beyond them, to join us. This simple idea involved Carrie coming to our house when they first arrived and we set a time. Then the next day, I snow-shoed down to Burt and Chris's cabin to invite them and then stopped at Warren's on the way back to confirm the time and the place. That night, we loaded our food into backpacks and the soup onto a sled and hiked down the ice where we enjoyed a wonderful evening of good food and laughter. Afterwards, we loaded up our leftovers and hiked back again, only to realize as we approached our cabin that I had forgotten my ski pole. So the next morning, I had to hike back again and fetch it.

Peggy Herbert

    Whether we are cooking over an open fire or eating out or cooking on the grill or on the gas stove, the one thing that is common is friends and neighbors who enjoy our island eating adventures usually surround us.

8 ~ What We Do When There's Nowhere to Go

THE IDEA OF BEING ON an island for any length of time would send some people into a panic. We frequently are asked what we do while we are there. We can't easily jump in the car to go shopping, there are not easy places to hike, and there is no TV to watch and no job to go to. Somehow, we all find things that we each enjoy doing, sometimes together and sometimes alone.

Of course, swimming tops the list. Many of us love splashing around on a hot day. Tom claims he goes swimming, but actually all he does is jump in, paddle around, and get out. It is always fun when we can drag him into the water for a rousing king-of-the-raft fight or a swim race. He does hold the record for morning dips in the lake no matter what the temperature. He has been known to jump in late fall and early spring, when the water temperature hovers around forty degrees. This is his bath. He dives in off the end of the dock, quickly gets back out and lathers up with Dr. Bronner's biodegradable soap. Then it is back in again to rinse off and out. If the weather is warm, he might swim out to the swim raft and back before heading in to make the coffee.

Tom jumping into the lake.

I use swimming as my exercise on the island. I put on my snorkel and start to swim along the shore using the large underwater boulders as my guide. I try and stay along the line where the boulders end and the sandy bottom begins. It is always interesting to see and recognize favorite rocks or see bottom feeding fish or a crayfish scuttling along. Lindsey and Kelly are more adventuresome, and like to swim across to the mainland and back. They always take someone in a kayak along as a spotter, just in case a boat comes down the lake.

Holly and I like to "noodle." Noodles are four-foot long foam tubes about two inches in diameter. They were developed as a swimming toy for kids. One summer my nieces and nephew were visiting, so I bought six of them. I also bought connectors for them. Soon Holly and I began to play around with the noodles and started learning all these fun things you could do with them.

First of all, they are impossible to sink. One little noodle will support a lot of weight. With two or three of them, I could make an unsinkable raft to float around the lake in. Holly was a non-swimmer, so she put them around her waist and practiced her swimming strokes.

But that was only the beginning. Soon we were exercising with them. We could hold out our arms with a noodle in each one and push them down to our sides through the water. I could hold one out in front of me and kick my legs really hard. I also learned that if I put one around me and another under my arms, I could bicycle my legs and begin to move through the water, just as if I were on a bicycle.

I also tried to balance on one. I would make two circles and put a foot into each one and then try to stand. It wouldn't be long before I was tipped over and went splashing into the water.

The funniest thing I saw was when my nephew Charlie Herbert brought his dog to the lake. Canella was a small sheltie that loved the water and particularly loved to chase things that you'd throw for her. One day when Charlie was swimming, Canella joined him. Charlie threw the noodle and Canella swam for it. Suddenly she had her paws over it and it was holding her up! Her back feet kept kicking so she kept moving through the water.

Another favorite activity is kayaking. We started out using canoes and still have two of them, but we have all come to prefer kayaks. We first bought an old double Folboat from Tom's cousin, but it turned out to be too heavy and too unwieldy, so we sold it and bought individual ones. I initially had a small Otter that I quickly realized was not quite what I wanted. It seemed to take a lot of work to keep it on track, and I was always lagging behind Tom. When I retired, I got a new, longer one that I love. The grandkids have even gotten into kayaking and have made surprising progress with the big kayaks with long paddles. So, they now have their own kid kayak with lower sides and a much shorter paddle to use.

Peggy Herbert

Kids in their kayaks

A few years ago we bought a 1950s vintage sailboat from our island neighbors. It is fiberglass and fairly seaworthy or so we thought. A couple of years ago, we were home in New Hampshire when we got a call from friends down the lake saying the sailboat had sunk and they had pulled it to shore. So on our next trip up, Tom and Todd sealed it up, and now if floats without taking on any water. Todd especially is the sailor of the group and loves going out with Kelly and the kids for a sail around the island.

Another favorite pastime now that the kids have gotten a bit older is to go tubing. Both Emma and Cooper love the thrill of sitting behind the big boat and being pulled around the lake. Seeing their fearlessness is awe inspiring and even when Emma goes flying off into the water, she just bobs up again laughing.

There are a lot of quiet activities to do. Holly, Diana and I laughingly have claimed to guard the lake, which means we play cards, read, do puzzles, knit or just talk. When I was at the lake alone for a three-week solo, I used my treadle sewing machine to make curtains. While some of us play, other folks find physical work to be a wonderful counterpoint to everyday jobs that take a lot of mental work. Tom is most happy when he can be cutting trails, hauling wood or building a new addition. Lenny and LeRoy, our most frequent guests, are always willing to help him with this.

Kelly and Emma reading on the deck.

We now all have t-shirts or sweatshirts that highlight what we like doing. LeRoy started this by having sweatshirts made for Tom and him that said *Mooselook Tree Service*. Tom followed up with t-shirts that said, *Mooselook Movers: Rocks and Trees Our Specialties--Division of Mooselook Tree Service*. The women were feeling left out so we made our own t-shirts that said, *Mooselook Meditators--What we do best is read and rest*. The most recent addition is new sweatshirts for the winter burn. These show a picture of an old-fashioned fire truck with the words *Mooselook Fire Department* written above it. LeRoy chose to have his oldest, grungiest sweatshirt embroidered, which fits in perfectly with this messy grungy job.

Obviously there is plenty to do out here in the middle of nowhere. But for those who are willing to leave the island, there are hikes, visits to local museums and shopping expeditions and restaurants. Kelly and Todd have made hiking Bald Mountain an annual tradition. We pack a lunch and numerous snacks and drive to the base and begin the climb. The books always say it is an easy hike. I, however, disagree. While the beginning is a gradual uphill on a well-worn path, the last section is steep with a lot of scrambling over huge rocks. I am usually exhausted by the time I reach the top, but I have to say the view is worth it. Bald Mountain is indeed bald and you can see Rangeley Lake, Mooselookmeguntic and Richardson once you climb the fire tower. It is a goal of the grandkids to walk farther on their own each year. By age four, both Cooper and Emma could go up and back. You may ask what that says about their grandmother…

But in spite of the off-island opportunities, generally once folks get on the island, they just don't want to leave.

Todd and the kids on top of the Bald Mountain fire tower.

9 ~ Tom's Trails

One of Tom's most constant activities has been the cutting of trails. It all started early on when I said that we needed someplace to walk. We had the area between the cabin and the lakeshore pretty well tamed. But you can only walk the seventy-five feet to the shore and back so many times. Little did I realize (although

I should have, knowing Tom), that my comment would lead to a vast interlocking network of trails that are the envy of our neighbors.

In 1989, we cut the first trail. It was the Shore Trail, which went east along the shore connecting us to the cabins down the lake. This one was fairly straightforward because we knew where it would go. It was just a matter of clearing the dead trees and branches out of the way and trying to make it as level as we could.

The next trails started up the east and west boundaries. These, too, were easy, as surveyors had painted property blazes onto the trees. The eastern boundary goes straight back and separates us from our neighbors. The western boundary is also the county line, and surveyors have come through every year or two to make sure it is well marked. It is much longer, due to the fact that it goes at a diagonal, and our four hundred feet of frontage becomes three thousand feet at the back boundary.

While we were cutting towards the back, Tom was also cutting cross trails that connected the two boundary trails at various points. It is these meandering paths that he really enjoys creating.

He starts by going out in the winter on his snowshoes. Winter is a good time, because there are no leaves on the trees and he can see through the woods more easily. Also, with the snow covering the blow-downs, it is easier to travel. He has a general idea of where he wants to go, so he heads out with bright orange surveyor's tape. As he walks through the woods, he looks for landmarks. These can be especially large birch or white pine trees, maybe an interesting rock or a clump of maples growing together. Going in a straight line is not the point. He flags branches along the way, and when he comes out to where he wants to be, that part of the job is over until summer when the grueling work begins.

When it is time to start, Tom dresses in his oldest clothes, gets his chain saw, safety helmet and chaps, some gas and oil, and heads to the beginning of the trail and starts cutting. He is never sure what he will find. Often, because of the winter snow depths, the tapes are high in the trees. And again, because of what the snow has covered, he may find that he has blazed a trail that goes through or over huge blow downs or large rocks. He modifies the trail as he goes along not only for ease of cutting, but also to avoid having to destroy healthy white pines or maple trees.

The usual procedure for trail clearing is to first take the chain saw and cut up downed trees and toss them to the side. He does this for about fifty feet, and then turns off the saw, turns around and walks back, trimming branches off trees with an ax. Then he goes back up to where he left his saw, clearing debris as he goes. He continues this until he has used up one full tank of gas. By that time, he is sweaty, dirty and tired, and ready to come back to the cabin and jump into the lake.

One of the biggest undertakings was to cut the back property line. This was not only the longest trail we had, but also the hardest to follow. It had been blazed long before we owned the land and the orange paint was faded or missing. We had walked it once right after we bought the property, and it was like crawling through a giant pile of pick-up sticks. Tom had done some work on it, but it seemed endless. We decided that to celebrate our tenth anniversary on the island, we would invite our friends to come and help us with this project. We divided into two teams. One team went with Tom and started at the west end of the trail. I took the other group and we started where Tom had quit cutting by himself at the eastern side. It was a hot, humid day, and by the time we met in the middle, we looked like members of a prison chain gang. Sweat was dripping through the dust that was covering our faces and we could barely walk we were so tired. We drove a golden spike into the ground, took our picture and headed back to the cabin and jumped into the lake.

Peggy Herbert

Tom, Lindsey, LeRoy, Diana, and friends finishing the trail.

While many of our trails go around huge trees, giant boulders or huge uprooted stumps, there is one that Tom has tried to keep as straight as possible. This is what he calls the "skidder" trail. It is in the middle of the property, and is used in the winter to drag hardwood out by sled.

We now have a series of trails all around and through our fifty acres. Of course the work is not done because they have to be maintained. This involves hiking all of them in the spring to see what has fallen across each trail. Sometimes all we have to do is throw stuff to the side, and other times it means Tom needs his chainsaw. The trail system is so intricate that we have drawn a map and made copies for people. Holly and Lenny have made and posted signs at every intersection. Not only are the trails all named, Tom has marked each trail with its own colored blaze markings. Kelly and Todd have used their GPS watch to begin mapping them. Our goal is to be able to print an accurate map from the computer.

While making the trails is obviously a favorite pastime for Tom, many of us enjoy walking on them. We can see how the land changes as we move further inland. While the shore frontage is primarily spruce and balsam fir, there are huge stands of maple further back. There are birch trees everywhere. We particularly love the huge yellow birch, which are so big that you cannot wrap your arms around them. To walk by one and realize that it was probably there during the Civil War (as we were told by a forester who was surveying our land) reinforces the forest mystique that so many of us feel.

10 ~ Boats….An Island Necessity

If any one thing has come close to making me want to give up owning part of an island, it has been our experiences with boats. Obviously a good boat is imperative. Our early canoe trips had proved that to us.

Our first few visits were by canoe, a borrowed one and ours. We would put in at Indian Cove, a beach that was a mile east of the cabin, and paddle across. It was on one of these visits that first fall when we realized we had to get a boat with a motor.

We woke to a bright sunny October Sunday. As I stared out from our tent, I could see the clouds scudding across the sky. It was a gorgeous day. I expected equal enthusiasm from Tom, and was surprised to see a frown on his face.

He told me to look at the lake. I did…and it was magnificent. Deep blue, wind ripples and here and there a white cap.

However, during the night, the wind, which we were prepared to have blow us in our canoes back to the landing and our car, had shifted and was now blowing in the opposite direction. This meant paddling into it as well as getting across to the other side, a prospect neither Tom in a canoe alone, nor I paddling with 11-year-old Lindsey, looked forward to.

We kept our senses tuned to the wind and became increasingly uneasy, as it did not abate. If anything, it became stronger. Finally, about 10:30, we decided to pack up and head across.

You would think that three experienced campers could spend two days camping with a minimum of gear. But between the sleeping bags, tent, chain saw and food supplies, we had quite a load. We began to regret these excesses as we started putting it all into the canoes.

Tom took off first. Lindsey and I wedged ourselves in amongst various pieces of equipment and started out…both feeling worried, as neither of us was a strong canoeist. We immediately realized just how challenging this trip was going to be. I kept yelling at Lindsey to paddle harder in an effort to keep the canoe heading into the waves rather than get turned sideways. If that happened, we could easily be overturned.

My arms began to ache. Every time I let up even the slightest bit, the canoe began to drift sideways. We watched Tom, who was making an equally gallant effort to keep his canoe headed forward. It became obvious that there was no way we would make it to the landing. Our goal became keeping the canoe straight and getting as far up the shore toward it as we could.

I found myself becoming furious, with the wind, with Tom, with myself for being in such an awful situation. With each stroke, I got madder and madder. I didn't want to be out there risking my life! But I also realized that I had no choice. I had to just keep paddling.

I saw that Tom had begun to head for the beach in front of the nearest cabin, partly by choice and partly because that is where the wind was taking him. I yelled to Lindsey that we would head for the same place. But no matter how hard we paddled, we couldn't get that far. The wind was pushing us into the shore along the rocks west of the cabin. At this point, I didn't care. All I wanted was to be in shallow water and know that we were not going to capsize.

As we got in waist deep water, I decided that the best thing to do was jump overboard and then lead the canoe out and around the rocks to the beach area where Tom was. The water was icy cold but I still loved the feel of solid ground under my feet. I waded to the front of the canoe, grabbed the rope and started pulling. I had to watch my step, as there were huge boulders underneath the water as well as those that we could see sticking out. Slowly we made our way until at last we got to the sandy area where we could get out.

Tom went and got the car while Lindsey and I hauled up our gear. We loaded up and the three of us piled in. As we drove off, we heaved a collective sigh of relief and agreed that our first expense for the next year would be a boat with a real motor.

Tales from Toothaker

So the winter of 1985-86 was spent combing the want ads hoping to find the perfect boat…sturdy, dependable and cheap.

In March we thought we had found it, a gold tri-hull. It was with excitement that we launched her in the lake during April vacation, a mere four days after ice-out. Because this was a "real boat," we needed a "real" landing, which meant Bemis Landing, three miles east at the end of the lake. After filling the boat with all our camping gear, we took off, feeling like we finally had some control over the lake. It took about two minutes to disabuse us of that notion as smoke began pouring out of the engine. We headed back to shore, loaded her onto the trailer and took her into Oquossoc Marine, where she was destined to spend most of her summer. We would get her back only to find new things going wrong that would prove difficult to fix. We used her approximately twice that summer, and even those times were shaky. Finally, at the end of the season, we towed her back to Concord, where she became a lawn ornament for the next year, her only decoration a fluorescent orange FOR SALE sign.

During all this time, we still needed to get to the island. Larry Koob, the owner of Oquossoc Marine, generously loaned us a variety of beaten up fishing boats. We bought a blue twelve foot one for $200 and then spend $1000 on a new nine-horse power motor. Finally we had something reliable.

However, we still needed a larger boat. We were back to watching the want ads and following leads. All winter Tom had nursed along a possibility of one down on the Cape, but when we saw her, we decided it wasn't for us. He then heard about a twenty-five year old aluminum Starcraft north of Bangor. After some phone calls and sending a friend to take pictures, we decided to buy it. Larry Koob had a fairly new seventy-horse power engine, so we thought we were set.

Of course now that we had a workable boat, the first thing we wanted to do was water-ski. So I sent Tom into Oquossoc to buy a life belt and towrope to go with some hand-me-down skies. Tom Lajoie offered to be our first skier.

We pulled him up and proceeded to whip him around in tight fast circles. I was the lookout ready to relay any signals. Suddenly I felt the boat slow down and watched Tom's skies slowly sinking into the water. It turned out that it wasn't my husband playing, but rather a broken steering cable that meant the boat no long went in any direction except circles. So the two Toms nursed the boat to Bemis landing and then took it into Oquossoc Marine, where it was quickly repaired.

Another problem happened our next trip up that season. Tom had come in one car with Lindsey and a friend. Kelly and I would be coming up later. We arrived to find them futilely trying to start the Starcraft. We thought it might be flooded, so we sat and waited and then tried again. Still nothing. Kelly was especially frustrated, as her rare trips always seemed to coincide with a boat calamity.

I suggested swimming over to the island to get the fishing boat, but we were able to borrow a canoe. After hauling people and gear over, Tom once again made the trip into Oquossoc Marine. This time the problem was minor: something wasn't plugged in securely. He was mostly embarrassed, but did manage to salvage the trip by buying a new bilge pump, which we needed.

Coming over during the day is one thing. Coming at night is a very different challenge. If there is a moon, we can see the island. But without one, it is so black that the island is invisible. The fall after we built the cabin, we learned just how dark it could be.

We were launching the fishing boat at Indian Cove when Tom suddenly announced that he had forgotten a flashlight. It was one of those very black nights with not a star or a sliver of moon to guide us. How would we find the cabin when we couldn't even find the island?

The four of us set off into the dark, Tom and I and Lindsey and a friend. We made our way carefully out of the cove. Once we were in the lake, I looked toward where I thought this 750-acre island should be. Although only about a half mile away, it could not be seen. Tom headed in what he thought was the right direction while the three of us peered intently into the darkness. We knew that once we got close to the island, we would need to follow the shoreline for a ways and then head into shore. If we went in too soon, we'd hit one of the many boulders that dotted the edge of the lake, especially at this time of year.

"I think I see something," I yelled. We all looked intently and indeed the blackness seemed somehow different. We realized we were seeing the island. But we still didn't know where along the shore we were. We were all wishing for the missing flashlight, which would have picked up the reflectors Tom had put along the dock.

Finally Tom said, "This should be about it," and he headed toward shore. I felt panicky wondering if we would be anywhere near our cabin. I waited for the crash of our boat against one of those lurking boulders.

Slowly, out of the gloom, there appeared a patch that wasn't quite so dark. "Hey!" Tom shouted. "I think I can see our roof." Sure enough, as we drew closer to shore, the foil-covered rigid insulation began to shimmer in the darkness. At that instant, I knew we were okay. We headed into shore and the dock appeared ready to receive us once again.

You may have noticed that after the early paddling adventure, nowhere is there any mention of me driving a boat. That is because I didn't ever drive it. I knew it was important to learn, but I was terrified. The boats all seemed to have minds of their own. By this time, we had upgraded from our original $200 fishing boat to something a bit larger and without leaks. So I decided to start with that, as it was smaller. In retrospect, I have to say, it is much harder to drive than the big boat, but that is not how my mind was working at the time. Small seemed somehow safer, even if the steering never made any sense. It took a bit of practice, but before long, I was able to take the small boat out on my own.

Then it was time to learn how to drive the 18-foot, 70 horsepower MONSTER. Tom kept suggesting it and I kept pretending I didn't hear him. Then I heard the words that made the difference. He said he would only go on our new inflatable skibob if I drove the boat. The idea of pulling Tom behind the boat while he rode on this torpedo shaped water toy made me see driving the boat in a whole new light. He, who so rejoiced in providing unusual rides for others, was finally going to get an "unusual" ride of his own.

First I practiced driving with Tom along to give me pointers. I was nervous as I sat behind the steering wheel. I knew I would never remember which lever to pull and which to push. How would I ever manage to get away from the dock without hitting it?

Much to my surprise, the boat started right up. I put it in reverse and backed up without hitting a thing. Then it became more fun as I sped up and tried going in circles. When I felt comfortable, I headed back to shore to drop Tom off so he could get the skibob and life vest.

Then we headed out into the middle of the lake. He threw in the skibob and jumped in after it. I watched him struggle to get on top of it, and when he finally yelled that he was ready, I took off. At first I went slow and let him get used to the ride. Pretty soon he decided he wanted to try and stand. This is what I had been waiting for. I let him try it a few times. Then, just as he was feeling pretty sure of himself, I pushed the throttle forward and the boat leapt ahead and off flew Tom into the water with a splash. He came up laughing and shaking his fist at me.

I picked him up and we headed for shore. Even though I knew Tom would now be waiting to get back at me, it had been worth it. I realized that driving the boat had not been so bad after all; it was like a lot of things in life that seem scary at first, but with a little practice are easy and fun.

One of the scariest crossings happened to Tom and his brother Dave. It was November of the year we built the cabin and they had arrived to finish putting down some floorboards and bring home the generator we had borrowed for the summer. They were in the small fishing boat with a strong westerly wind blowing at their backs, creating three to four foot waves. They were about half way across when the motor quit. As Tom pulled on the cord to restart it, the cord came off in his hand. They picked up the oars. Not only were they covered in ice, but they also kept popping out of the oarlocks. A couple of huge waves appeared, but they successfully rode over them. These were followed by a couple more huge waves that they didn't ride over. Instead they broke over the back of the boat soaking them. They kept rowing, as it seemed to be their only option. Finally they made it to shore and gratefully pulled out the boat and happily secured it for the winter.

Tom's nephew Charlie, along with friends, did our favorite lake crossing. They wanted to visit the cabin when we were not there. but were not able to find a canoe or boat to use. When he talked about swimming, we became nervous due to the rough water and the difficulty of being seen by motorized boats.

They solved this problem by using an inflatable float for emergency flotation and a Big Bird helium balloon as a visual warning. After a successful crossing, they took our canoe back for their gear. They reversed this procedure when they were ready to leave.

Tales from Toothaker

There is another part of the boat story that, while it isn't connected directly to using the boat in the water, is critical to being able to use the boat at all. This is the ability to take the boat in and out of the lake

For the first ten years we had been going up to Lake Mooselookmeguntic, I had never dared stay there alone. Not because I was afraid of the loneliness, or the wildlife or the isolation. The simple reason was that I was too chicken to back up the boat trailer to get the boat in and out of the water! However, an opportunity presented itself that I couldn't resist. We had two cars at the landing, and while Tom needed to get home, there was no reason that I couldn't stay a couple of days longer. This meant I would only have pull the boat out and then park it. I wanted to stay by myself badly enough to risk this challenge.

As the time to leave drew near, I began to get nervous and started remembering my one earlier attempt to back the trailer down the ramp. It sounds like an easy thing to do, but I had learned it is not. It never seemed to go where I wanted it to go. When I turned the truck the way that seemed logical, the trailer suddenly veered off in a totally unexpected direction. Despite repeated attempts, it was clearly the winner that first time.

This time, I was going to be the winner. Tom had prepared me before he left by drawing little diagrams and using forks and spoons to try and explain how I had to turn the truck to get the trailer going in the right direction.

I had a wonderful relaxing stay, and finally it was time to leave. I headed off with Luke, our dog at the time, as my first mate. We made our way down the lake towards the landing at Bemis, me in back at the motor and Luke up front, ears flying and tail wagging.

The first thing I needed to do was land the boat at the dock. That part didn't bother me--until I remembered all those new suggestions Tom had shared about how to avoid the big rock that lay in wait to snag unsuspecting motors. In the middle of trying to make my numerous rights and lefts, I hit smack dab into the end of the dock. Luke jumped out while I killed the motor and ran to the front of the boat. The water did not appear deep, so I jumped in--only to find it was up to my waist. I looked toward the cabin hundred feet away and hoped no one was watching.

I tied the boat up, ran up to the parking lot where I got the trailer hitched to the truck and drove it into position to back up. I needed to get the trailer to turn ninety degrees from the driveway and head down an eight-foot wide concrete ramp into the lake.

On my first attempt, I got the trailer onto the ramp; but instead of heading down the ramp, it was going crossways and was about to fall off the other side. I kept trying to back up, but the only thing that was happening was that I was getting hot and impatient. On my next try, when it still wasn't lined up correctly, I decided I'd get out of the truck and move the trailer by hand. I gave up on that idea when I couldn't budge it. So I pulled the truck up again and this time, through some unexplainable miracle, I got it lined up. I couldn't believe it. I carefully backed down, keeping it perfectly straight. I loaded the boat on and felt like a pro as I pulled it up and out of the water.

I drove back to the parking area where we left the boat between visits. I hoped this time backing up would be easier as it is a huge lot--but it wasn't. It took more repeated frustrating attempts, until at last, another miracle and the boat ended up just where I wanted it. Now all I had to do was unhitch the trailer and head home.

I jumped out of the truck, positioned the block of wood we rested the trailer on, flipped the lever that held the ball and lifted. Nothing happened. I yanked and pulled…still the ball held fast. I pounded and whacked. I jumped on the back of the boat. I couldn't believe that I had gotten this far and wasn't able to get the trailer free from the truck.

It was then that the third miracle occurred. I reached down for one more try, and the hitch effortlessly lifted off the ball. I set the trailer on the block, called Luke and started off for home. I looked out my rear view mirror at the boat sitting right where it belonged and thought about how often I want things to happen quickly and effortlessly. The boat was a reminder once again, that patience and practice are often more important tools than speed and ease to getting a job done right.

When we bought our land, we also had a lease that allowed us to launch our boat at Bemis as well as leave it down there between visits. This worked well, although it was a long ways away, and getting in and out in low water could be difficult. It was also hard to know when people arrived as it we couldn't see them and we didn't have cell phone service. When we started hearing rumors of a boat launch across from us on state land we had mixed feelings. Our big fear was that our lake would suddenly be filled with boaters and noise. When rumor became reality and the state actually put the launch and huge parking lot in directly across from us, we realized our fears were unfounded. In fact, it has been a godsend, allowing us easy access, a much shorter boat ride and the ability to see when people arrive.

We still have our aluminum fishing boat and the 9.9 motor. But after eighteen years of using our dependable 1964 Starcraft, Oquossoc Marine told us that it was unsafe and unrepairable. A hunt for a new boat began. The deep hull of the Starcraft worked well on the lake, so we decided to see if we could find a similar model. It was much easier this time as we now had access to the Internet. Tom went onto E-Bay, typed in what he wanted, and started going through the results. He found just the boat he was looking for in Pennsylvania, so he spent a weekend driving down, picking it up and returning home.

We also have a small, very old sailboat, four kayaks and two canoes. And up against some trees is the old blue fishing boat that we started with. So, unlike those early days, we now always have some means of being on the water whatever the conditions.

11 ~ Ever-Evolving Dock Systems

One of the ongoing projects is planning, locating, repairing and putting in docks. Most of the other cabins have huge boulders all along the shoreline that are more or less exposed, depending on the level of the lake. We have one huge rock and a few medium-sized ones, which means we have a lot more latitude in where we can set our dock.

Within the first year, I located what I still think is the best spot along the shore. It drops off quickly into deep water and is fairly clear of rocks. In order to get to it however, we had to clear a 30-foot path through the scrub spruce trees. Once this was done, it was an easy walk up to the cabin.

When we first started, we did not realize the rules about no permanent docks, and so Tom was very keen on building a crib dock like his brother Pete's in northern Minnesota. Pete sent elaborate plans that involved building a 4x4 foot crib out of logs held together with metal "ready rods" in the four corners. This was placed a ways out from shore and filled with rocks, the bigger the better. Then long log stringers were run out from shore, set on the rods, and planking set across them to form the dock. Obviously this was only the beginning, because we also needed some floating docks that could drop down during the summer when the dam released water and the lake level dropped. So Tom built two three-by-eight foot sections that he put flotation under, and put those out beyond the permanent dock.

Then the question became one of how to get from the permanent dock down to the floating docks as the water dropped. Tom took care of this with a ramp that he attached in such a way that one end would lower, creating a walkway that would get steeper and steeper as the summer went on and the power company pulled more water through the dam.

This dock system worked great except for one thing…what the ice did to it each winter. The two floating sections were removed in the fall, but the permanent structure stayed, almost completely out of the water by October. However, by spring break-up, the water was once again high and filled with pounding slabs of ice that had the power to twist and break our crib. In fact, one spring, after ice out, Warren called us to say he had boated past our place and the dock "didn't look quite right." When Tom and Lindsey went up for the first time, they found the ice had driven the crib with the stringers up into the woods. They called Warren and told him he was absolutely right, something was just not "quite right" about the dock. We decided we needed a new system. By this time, we also realized that permanent docks were prohibited on the island, so that helped Tom decide what to do next.

The system that emerged was very similar to the last one, except that what used to be permanent was now a huge heavy piece of docking that was pulled up in the fall and pushed back in the spring. Initially this dock rested on a smaller removable crib and a large plastic barrel. The two floating docks were then attached. As time went on, we replaced the barrel and crib with two pipes at the outside corners. The ramp to the floating sections would be adjusted up or down depending on water levels.

Putting in the dock each spring is always a challenge. Ice goes out on the lake anywhere from late April to mid-May. It is generally Memorial Day when we arrive to open up. All this really means is that we take out the deck furniture, put away the wood stove and put in the dock. Nowadays, putting in the dock involves standing in freezing cold water in wet suits trying to maneuver two hundred pound pieces of docking down into place to be secured in the water. I have only had to do this once, as we usually have some male who is a lot stronger and enjoys the challenge more than I do. In recent years, this has been Lenny.

Tom and Lenny putting in the dock

The first year Lenny did this was especially memorable. We got out the hand-me-down wet suits and he and Tom each got one. Tom pulled on his one-piece over- the-shoulder leggings while cracking all kinds of silly jokes. Lenny, meanwhile, was sitting to the side slowly struggling, trying to get his bottoms pulled up to his waist. He was panting and straining--it reminded me of putting on a girdle. Holly and I sat by helplessly; as there is really no way you can help out in a situation like this. Finally with one last pull, he got them up and proudly stood. Although his face was still red, his breathing had begun to slow down as he bent over to zip up the ankle zipper on the legs. He was suddenly calling for help, as he couldn't find the pull on the zipper. Holly went over and she couldn't find it either. As we all stared at it, it suddenly hit us that the reason we couldn't find it is because it was on the inside of his pants…which meant that after all that struggle, he had ended up with his bottoms on inside out!

He fell back onto the couch with a groan while Holly grabbed hold and tried to pull them off. This turned out to be as much work as putting them on. Finally with a snap, they flew off and poor Lenny had to start all over again. Eventually he was dressed, and he and Tom were ready to start outside…only to find it was snowing. They were not going to let this stop them, and so on they went, leaving footprints in the snow as they made their way toward the water.

Holly helping Lenny get into a wetsuit

Our most recent innovation occurred because of unusually high water one summer. In an ordinary year, the water drops and exposes a beach area where we can sit and the kids can swim. When this spot never appeared, we found that we were spending much of our time on the dock while the grandkids splashed in the water nearby. Consequently, we were always bumping into someone or running out of room for chairs as we jockeyed for space on the five-foot wide dock. We clearly needed more room. As I looked out at the 8'x8' swim raft, I wondered if something like that could be added to the end of the existing dock system. Tom was initially opposed to the plan. He saw it as one more thing to put in and take out. But being the good sport that he is, he decided to give it a try and initially modified the swim raft to fit off the end of the current dock. Everyone loved it…even Tom. As far as I was concerned the project was done. However, unbeknownst to me, Tom had other ideas in mind. Here is his version of what happened next.

We were at the lake. Lenny had come up to help me pull out the dock. Peg left on Saturday morning. As we tend to do, Lenny and I started talking. Projects came on the agenda. The dock addition came up. Lenny said that this was the time to do it, when the water was low. His point was that once it was built it could not be moved, so it should be built right on the beach. I pointed out that the raft/extension would have to be flipped upside down in order to attach the boards to hold in the flotation billets. So we talked about how that could be accomplished. Then I pointed out that few of those ideas could be accomplished by just one person, working alone. So we talked some more and then dropped it.

Sunday night I decided I would just go ahead and build it and I would figure out how to flip it when I got to that point. On Monday morning I went to town, got what I needed, was back at the boat ramp by 9:30, and had everything hauled over to the cabin by 10. By lunchtime, the frame was built and the dividers were installed. We were now at "that point" where it needed to be flipped.

I put a sling up in a tree, with the help of an extension ladder. Attached to the sling were two alloy carabineers. Through the carabineers I ran a piece of goldline climbing rope, one end attached to the raft frame, the other to a come-along, which I had anchored to a huge boulder with an eyebolt in it. I cranked, and just what I expected to happen, happened. Nothing. The safety stretch that is built into the climbing ropes, did what it was designed to do….stretch.

I replaced the rope with cable and the two carabineers with two steel chain links. I cranked on the come-along again. This time it lifted up the end. I kept cranking until it was almost vertical. I thought next about attaching the stringers…better to use the drill and sheet rock screws for the ones way under so there would be less jarring. Didn't want this thing to fall on top of me!

Another thought…I had just finished writing the emergency protocol for the cabin…what to do in case of serious accidents. This had the potential for one of those serious accidents…plus I was there by myself. If this 300-pound raft fell on top of me, there was no way to get help. I was the only one on this side of the island and I had not seen one boat all day. So, I put a back-up chain around the tree and affixed it to the raft frame. Then I attached two legs to the front and reinforced those. This baby was going no place unless I wanted it to!

I put the stringers on, I lowered the raft down, I put in the billets. The next day I decked it, put on the connectors, put on the weights and the ladder attachment, and finished it off with the reflectors. It was ready to launch.

Peggy Herbert

I knew nothing of all this. Tom came home and didn't say a thing. It was a few weeks later, just before my birthday when we next went to the cabin. As we went across the lake, Tom told me to turn around and face him and he handed me an envelope. We landed the boat and he told me I could finally look. There was the new dock. I opened the envelope to find the above story as well as pictures. And he ended with this: *So happy birthday. You have a dock extension…not 8'x8' but 8'x10'… ready to go in the spring, as soon as ice is out.*

Tom, Brenda, Kelly and Beth Ann enjoying the new dock extension.

12 ~ Our "Neighborhood"

When we first bought our land in 1985, there were no cabins at all on our side of the island. The primitiveness of it all never struck me more than the time Lindsey and her friend Sarah Anderson took out the fishing boat using only the oars. All of us being novices, we didn't stop to think that the prevailing wind that was helping push them down the lake was going to make it impossible for two twelve-year olds to row back. We could see them at a distance struggling valiantly, but making no progress. Tom decided to head down through the woods and try to meet up with them. I remember looking at those endless woods and feeling extremely isolated and alone. I could hear him crashing through the blow-down as he worked his way down the shore. He finally met up with the girls and I felt my panic subside.

I often think of that incident as I look at that same stretch of woods now. In the years since, a number of people have bought land and built cabins. One of them actually cut the logs off his land and used them to build his cabin. Watching the amount of work that went into this convinced the rest of us to order our logs precut.

There are now twenty-two cabins on the island…all inhabited with people who have a certain necessary island mentality. Although in many ways we are very different, we all are willing to work hard and put up with certain hardships in order to enjoy what life has to offer us up here in the wilderness.

The first residents Heidi Sorenson and Larry Hall bought a huge parcel next to the county line on the north side of the island. For many years they lived there full time. Also on that side, in the mid-1980s, the Graham family bought adjoining acreage and gradually built five camps, where each year island residents gather for the annual 4th of July potluck.

In 1985, parcels on the south side of the island were sold to another extended family, John and Priscilla McAdams, John's sister Sue and her husband Brian Dougal and their nieces Debbie and Jan and their husbands Rich Emmons and Dave Ahern. They now have four cabins and host the end of the summer gathering over Labor Day.

At this same time, we bought our piece of property. Shelton Noyes told us that three military friends had purchased three pieces on our side, sight unseen. Apparently the man who had bought ours had driven up here from the south with his wife. They launched their boat and came out to the island. His wife took one look at the wild tangle she found and told her husband to sell it.

Shortly after we bought, other pieces began to be sold until we had quite a community. A few folks have bought and then sold after realizing the challenges of island life. In fact one family built a beautiful large home on the east end of the island using logs that they had harvested from their land. Their dream was to live in the house year round and home-school their four boys. This never quite worked out the way they had envisioned, and the house was sold to someone who uses it seasonally.

A few years ago a man bought a lot with the idea of building a real house. He wanted to know if the rest of us would be willing to chip in to have electricity run out here. There was a universal no to his request. So he went ahead and built his house, but uses a generator to power it.

One of the first things we did was clear a trail along the shore to get from one camp to another. Just having that trail took away a lot of the early feeling of vast isolation I had felt when Lindsey and Sarah got stranded. And it is certainly nice to be able to have neighbors to borrow from, as it is a long ways into Rangeley to get what you need. We have borrowed and lent things like burn salve, saws, a drawknife or a battery to jump a dead car. My favorite was the time we went down to one neighbor to "borrow" two twelve-foot long boards that we needed to finish a project. They happened to have them and were glad to "lend" them. We replaced them the next time we came up.

Some people we see more often than others…just because of proximity or interests. We share books and food with the Kitzmiller clan and visits and trail clearing with Burt and Chris. Each season it is fun to catch up on what everyone has been up to during the past year. We sit around the campfire and hear tales of children graduating, jobs changing or vacations taken.

Even though we don't see our island friends often away from the lake, as time goes on, we realize how deeply our connections go with some of them. When Warren's brother-in-law Kip died of a sudden heart attack, we were devastated. And when Warren's wife (and Kip's sister) died of cancer four months later, we felt like we had lost two of the people whose presence at the lake had always enriched the experience for the rest of us. Through their deaths, we realize that as idyllic as our island life may seem, we are not immune to the losses we all face no matter where we are. And how glad we are to share our island life with our island neighbors.

13 ~ Guests, Expected and the Unexpected

Living on an island means that drop-in guests are virtually unheard of. Guests have a well-planned trip with arrival times set so we will be ready to meet them with the boat at the landing across the lake. Some people are regulars who come every year. Then there are those who live far away and are not able to come as often. We have had new babies and eighty-five year olds. My mom has come a couple of times, and, with her adventuresome spirit, loved it. She would come again at ninety if it weren't for the difficulty of getting in and out of the boat. We tried to get my dad up once, but having grown up with outhouses, he was not eager to repeat the experience. We have had visitors from Great Britain, New Zealand and France. However there have two rather notable "guests" who have shown up with no warning.

The first guest appeared during a winter visit soon after the cabin was built. The road was barely plowed, so we had to drive as far as we could and then hike over pulling our supplies.

Brenda and her husband Chris Dube were with us, and a night ski seemed like a good idea. Skies were donned and off we headed out onto the lake. It was cold, windy and very dark. Brenda and I decided the cabin sounded like a better place to be and the two of us turned around. We were heading back when a light appeared out of the darkness. Had Tom and Chris turned around already? The direction the glow was coming from did not seem to be where the two of them should be.

Moving back towards the cabin we kept our eye on the approaching light. Gradually a single shape appeared out of the gloom. Brenda, who lived in Boston and took no chances, immediately headed up onto the shore where she could hide in the trees. My reaction was to stay where I was and watch. As the figure drew closer, I saw it was a rather large man with a flashlight held in bare hands. He had no boots and no hat and was shaking with cold.

The last thing I expected was for him to ask to use our phone! He had gotten stuck on Upper Dam Road and the only light he had seen was ours. By this time Brenda had reappeared from the woods and the three of us headed into the cabin where the wood stove was going. Dismay shown on his face when he learned there was no phone. He had been looking for a friend's house, which, it turned out, was on a totally different road. He was stuck, he was lost and he was ill prepared for the weather. By this time Tom and Chris had returned and we offered to hike back across the lake with shovels and try to dig him out.

His car was quickly freed and off he went with new directions, while we hiked back laughing at what had just happened.

Our second surprise visitor was many years later. I had arrived at the island one Monday afternoon in late September, ready to begin my three-week post retirement solo. I had successfully launched the boat and managed to get it over to the island filled with topsoil, books, knitting, food, wine and other miscellaneous things that were essential for my stay. With the key in one hand and a duffle bag in the other, I headed up to the cabin to open up. I dropped the duffle in the kitchen and headed into the bedroom to get my sandals that were next to the far wall. What happened next is hard to remember…did I hear something or did my eyes just happen to look over? But I suddenly realized that there was a body between the bed and the wall. There was an instant flash that it was Tom, an idea that was just as quickly dismissed as impossible. Then I saw it was a teenage boy wearing a Muscular Dystrophy bike shirt so my next thought was it was someone Tom knew from his many years of riding for them. By this time the boy was standing and shaking so badly that he could hardly talk. I told him to take a few deep breaths and while he did, I studied him. He looked to be in his late teens. He was about six feet tall, well built with very light colored hair and eyebrows. When he began to speak, it was with a southern accent. The shirt was borrowed, he said because his was wet. As he said this, I flashed back to the damp shirt I had seen hanging on the railing by the beach. I'd ignored it at the time thinking it was just one someone had left at our last visit.

The story that finally emerged was that my intruder was attending Dynamy…an internship program for high school grads who do not want to start college right away. The initial part of the program is a wilderness experience run by Outward Bound. The group was on the island at the campsites and the students were all on their "solo" experiences. Neil (he had introduced himself) had seen the cabin, found the door unlocked so had gone in and gotten some chips and a drink. He couldn't stop apologizing--and wanted to know if there was anything he could do to help me. I quickly grabbed onto that and had him haul two hundred pounds of topsoil up to the garden. In my mind I am seeing a nice kid who had made a really stupid mistake. I told him it was up to him to tell the staff…but he didn't seem to want to do that…as I am sure he knew the consequence would be that he would be kicked out. I wished him well and sent him on his way.

While all this had been going on, Tom had called and left a message so I called him back. When I relayed my experience, he was dumbfounded, but I assured him it had all worked out fine and not to worry. Later that night there was yet another message telling me that Lenny--who had locked up on our last visit, knew that he had gone back and checked the door to make sure it was locked. My job was to check carefully and see if there was any other way Neil could have gotten in. I could see no sign of a break-in, although I did notice that the door to the shop was unlocked. Neil must have taken the key from the cabin to open that. But the cabin itself looked fine. I even looked at the two second floor end windows thinking maybe he'd used a ladder but they were secure as well. The only thing I noticed that was at all odd was one of the outside chairs was on top of the picnic table that is on the deck. I thought maybe Neil had put it up there to sit and get a better view of the lake.

It was a couple of days later when I was making curtains for the upstairs windows that I came across something odd…but even then I didn't associate it with the break-in. I was measuring the skylight in the bedroom when I saw a small tear in the screen. Odd I thought.

Maybe Lindsey had done it trying to close the window. The window was slightly open so I decided to close it and started cranking. No matter how I tried, I couldn't get the window to close tightly. I started jiggling it and soon saw the reason why: the plastic piece that connects the pulley to the actual window had been broken. You'd think at this point, things would start to click. It wasn't until the following day that all of the pieces fell into place. Maybe it was glancing at the chair that was still on the picnic table. But I suddenly knew how Neil had gotten in. The door *had* been locked and tightly closed. Neil, noticing that the skylight was slightly ajar, had put the chair up on the table, climbed onto the roof and broken the plastic piece, which had allowed him to lift the window. He slit the screen to remove it and climbed in…right onto the bed. He put everything back and made himself at home.

As far as I was concerned, this changed everything. I called Tom who had already had called Dynamy feeling that they ought to be alerted to what had happened. They were then updated with the new information, and Neil, who was still out with Outward Bound, was confronted and promptly burst into tears. He was dismissed from the program, his parents were contacted and he was to be sent home. We later learned that he had been in a residential school for his problems with impulse control! It seems he still needed some work.

People continue to ask me if I wasn't scared, and I always tell them no, for some reason I wasn't. Maybe it was Neil's shaking or his repeated apologies, but throughout the ordeal, I never thought to be afraid of him. But it still is not an experience I'd like to repeat. Given where we live, the planned-for visits with people we know are best.

14 ~ What to Be Afraid Of

Being on an isolated island is not without its dangers. We have been lucky that we have not had any serious medical emergencies. The wind, which offers great challenges to staying safe on the water, has scared us a few times, but we have never suffered any serious mishaps from it. We begin to take it all for granted until something happens to remind us of the precarious and treacherous nature of life on Lake Mooselookmeguntic.

A few years ago, a man drowned after he and his friend were both knocked out of their boat due to the strong winds that were causing rough waters. Neither of them was wearing a life jacket. His friend made it to safety, but the other fellow was not found for another few weeks when he finally drifted ashore on the other side of the lake. Until then, we were all a little leery of swimming, knowing that he was floating out there somewhere.

An incident that we witnessed luckily did not end in disaster, but could have easily done so. It was a windy day when we noticed some kids and their leaders loading their canoes and leaving the campsite up the island from us. We were concerned because the wind had picked up and white caps were rolling down the lake causing huge swells. The canoes were headed for the opposite shore and calmer water, but were being hit broadside by the waves. To make matters worse, no one was wearing a life jacket.

I kept my eyes on them while Tom went back to finish one of his projects. The first three canoes reached the far shore, but the remaining two were still fighting their way across. They were about in the middle when suddenly one canoe didn't look right. In the heavy waves, it was hard to tell if they had tipped over or were just in a trough. It didn't take long to realize that they had capsized. I had a pit in my stomach as I yelled for Tom. He dropped his tool belt where he stood and ran to the dock. We jumped in our big boat and headed out to the group. We knew there had been three people in each canoe and were relieved when we quickly counted three bodies struggling to put on life jackets and trying to gather floating paddles. The remaining canoe was staying right with them, but the help they could give was minimal, as it was all they could do to just keep themselves on an even keel. We got the floating campers into the boat and headed back to shore. We took them to our dock and went back to get the canoe. We sent the second canoe to our dock as well while we attached the overturned craft to the boat and pulled it to safety.

By this time, the remaining canoes were pulled up on the far side of the lake on a small beach. After gathering what floating debris we could from the water, we took both canoes and the six canoeists up to join the rest of their crew.

There they sat for the rest of the day as the wind kept up its blustering. At dusk we headed back down to where they were marooned and helped evacuate them to the mainland where they had started. One of the leaders gratefully presented us with an Old Town Canoe paddle as one of the campers announced that this was the best day of the trip.

More recently Tom, Kelly and Todd were down on the dock on another very windy day when they spied a 10-foot fishing boat coming across from the boat launch. Getting the binoculars, they realized the boat was carrying 6 people and none of them was wearing a lifejacket. Tom had just turned his head away when Kelly let out a yell. The boat had gone over. Tom and Todd jumped in our boat and headed out to rescue them. They found that the boat had flipped upside down and the six people were hanging onto the six inches of the boat that was above the water. The group was terrified. It turned out to be two women, two teenage boys, an 8-year-old girl and Alex, the male driver. With some careful maneuvering, they got everyone into our boat and took them back to the shore. Then they headed back to the partially submerged boat with Alex, who proceeded to jump out and try to flip it back over. When that didn't work, he threw a 15-pound anchor at our boat, which simply clanged off the side and fell into the water where it got stuck. Eventually his craft was rescued and brought back to shore, where Alex thought he would get it going and once again attempt to take everyone out to the island campsite where they were staying. While all this was going on, a neighbor appeared to help, only to run out of gas before he even reached the site. So besides hauling everyone out to the island, Tom also stopped to get a container of gas for the other boat.

The two women were not only scared but also totally embarrassed by what had happened. They realized that they should have known better. The next day when Tom went down to get them to take them back to their cars, they could not stop thanking him and wanted to pay for gas. Tom refused their offer and said that all he wanted was that they "pay it forward," and be there when someone else needed help. He was pleased when he got letters from one woman and her two sons thanking him and acknowledging their terror. They also wrote about what they had learned and said they had already signed up for a water safety course. I know they realized that had they not been seen, the consequences could have been deadly.

It is easy to sit back and watch other folks do things we think are foolish or dangerous without stopping to think how easy it is to get into our own precarious situation.

It was one Veteran's Day weekend when we decided to head north with Holly and Lenny. Wanting to have the most time possible, we chose to leave right after school, which would have us launching the boat in the dark. The fact that the temperature would be at freezing and the water level would have dropped to new lows didn't deter us. We had the spotlight to lead the way and Tom's good sense to drive the boat.

The trip to Maine went by quickly. It wasn't until we were coming up over the back road that we began to feel uneasy seeing snowflakes drifting down and the wind blowing the trees limbs back and forth. At Bemis, where we had left our boat turned over for winter, we piled out of the van to flip it back and put it on the trailer. But before we did anything, we all scrambled for hats, gloves, extra coats--anything to cut off the frigid wind. We hurriedly got the boat loaded and headed the last four miles to the landing.

Lenny got out and directed Tom as he backed down the ramp. We could see Lenny's arms flailing around and hear his yells as he tried to get the boat headed in a straight line. Our attention was particularly caught when he shouted to be careful, as the ramp was very slippery.

Tom kept backing and backing--the lake had dropped so much that the whole ramp was out of the water. He stopped short of the lake and we got out to unload. The wind seemed less and Lenny and I were hopeful that it was dropping. Holly, however, had the spotlight and was seeing the large whitecaps out in the middle.

We quickly put on our life vests, launched the boat, loaded all our gear, and then had to get free of the rocks. Tom and I both jumped in and sat shivering, waiting for Holly and Lenny. We looked back to see Lenny hunting to find a place to put his glasses in one of the forty-seven pockets in his new jacket. They finally got in and we pushed off. Tom and Lenny paddled out until we were in deep enough water to start the motor. We were all relieved to hear its hum when Tom pulled the starter cord.

It took about thirty seconds for me to realize that this trip was a big mistake. The waves were huge and the boat was rocking back and forth as we headed out across the water. I yelled at Tom to turn around but he just kept going. He in turn was yelling at Holly, who held the spotlight, as he needed her to find the reflector he had nailed to a tree that would show him where the island was. Holly in turn shouted at Lenny, who was right in front of her and blocking her view. He would then bend down and the spotlight would scan back and forth until suddenly the red reflector would appear and Tom would re-aim the boat. Meanwhile the waves were crashing over us and drenching us in freezing cold water. Tom would be watching the island when a huge wave would appear and hide everything. Then suddenly the light went out and everybody screamed. I had stepped on the cord and disconnected the light from the battery. When we had that fixed, we once again had to find the reflector that never seemed to be where we were expecting it.

I began to notice a pain in my thigh and realized that Holly was squeezing my leg. Every once in a while I would hear a kind of wild manic laugh come out of her mouth and I could never be sure if she thought this was fun or if she was terrified. Given the strength of her squeeze, I decided it was terror.

Through this, I was imagining what would happen if we went over. I was constantly trying to judge which shore we were closer to…should I swim back or onto the island. My relief was huge when we began to see island shoreline, and I knew then that we were safe no matter what.

It took awhile to maneuver around the rocks. As we got closer, Tom started stripping down to his underwear in preparation for jumping in. Lenny and I just jumped in…at this point stripping down seemed pointless. We got the boat pulled up and I ran to the cabin to start the fire. We were finally safe and beginning to once again get warm. Holly and I broke out the wine and wondered at our sanity.

When we mentioned our concerns to Tom and Lenny, they just laughed. Holly announced that she didn't think we should ever sit and laugh at the things other people did. Tom's response was that of course we could. We did our stupid thing in the dark so no one could see it. If people choose to do it in daylight, then they were fair game.

On Monday, we drove to Bemis where we stowed the boat for the winter. As we left, we met up with the carpenter who was building a couple of places on the island. Tom has met him before so he knew who we were. When he heard we had crossed on Friday night, he jumped out of his truck and started shaking our hands. He was amazed as they had tried to cross during the day and it was way too rough. The wind was 40-45 mph with gusts up to 60 with seven-foot waves. Another builder had tried to go out on a big raft and had almost been flipped end over end. Maybe if we had seen what we were actually getting into we would never have tried. But we made it and now have another good story to tell about living at the lake.

Having successfully survived our previous year's Veteran's Day crossing, we decided to head back up to the island once again for the long November weekend. Besides our usual four, Tom, Lenny, Holly and I, we were to be joined by our nephew Charlie and his fiancée Erin.

Tom and Lenny had been planning this trip for the past year. Their plans included warmer clothing, a possible motel stay if it was too rough to cross and most importantly, a new rechargeable high-powered spotlight that I could not inadvertently put out by stepping on a wire.

A sudden November snowstorm on the Wednesday before our trip caused further modifications of our plans. We now needed to think about how much snow might be in the boat, could we pull the boat out of the field where we store it and how slippery the boat ramp would be.

We decided that by adding a bucket of sand to our supplies, we could handle all of these new problems especially in light of a warming trend that was on the way.

We left home in bright sunshine on Friday afternoon. By the time we stopped for dinner in Gorham, it was pitch dark out and cold. When we got near the lake, we decided to try the little-used lumber company road and found that while it wasn't plowed and we tended to scrap our bottom, we could get still get through. Part way up we realized that there was a car ahead of us and discovered Charlie and Erin who quickly pulled over and let our van lead the way.

We arrived at the overgrown driveway that led to our boat. The plow had piled up snow in front of it but Lenny decided we could charge through and get the 100 feet in to the trailer. He opened the gate and sure enough, in we went. Everyone jumped out and started putting coats and hats and gloves… except for me who, remembering last year, had wisely put all my warm clothing on while we were driving.

We hooked up the boat trailer and discovered that without four-wheel drive, we did not have enough traction to pull the boat out through the snow, even with five of us pushing it. The van just kept sliding and spinning. Finally it was decided that we'd have to take the boat off the trailer. We still couldn't get out. So we unhitched the trailer and tried again. Still no luck. The van just kept spinning its wheels. We threw sand down and that, combined with all the pushing, finally got us back out onto the road.

Then we headed back and dragged the boat trailer out and hooked it up. Our final trip was to get the boat, which we carried out and settled back on the trailer. We were all standing there covered in snow when a car passed us pulling snowmobiles! The ridiculousness of the situation hit us as we laughingly got back into our vehicles and headed for the boat launch.

Once there, we realized that there was no way we could drive the car down the ramp and ever expect to get it back up again. So instead we took the boat off the trailer and turned it into a luge and slid it down, which turned out to be an easy and effortless way to get the boat into the water. All that remained to do was load up all of our gear and get across. The lake was fairly calm, there was enough light to see the island, so this turned out to be the easiest part of the night's adventures.

By our mid-afternoon departure on Monday, the temperature was a balmy sixty degrees and there was not a drop of snow left on the boat ramp. We got the boat pulled out and into its winter resting place and headed home already wondering what *next* November might bring in the way of challenges.

While many dangerous things happen on the water, there are also dangers on the island itself. Our biggest fear is fire.

Toothaker Island is a spruce forest with a few deciduous trees. The ground isn't really dirt, not like you find in a garden. What's there instead is about a foot of crumbly brown stuff that is made up of trees, leaves and pine needles that have slowly rotted and turned very dry. Because it is all organic material, it burns. If you start a campfire without putting down lots of sand and stones first, the fire will go right into the ground and start burning a path through the duff.

Fire can smolder under the ground for days and you might never know it. Then suddenly it can burn through, get new oxygen and start a big fire long after people have left the area.

The first time we experienced an underground fire was one of our first summers on the island. We were there with Lindsey and her friend Sarah. Tom had gone down to Bemis Landing to bring a load of equipment over. When he got back, he commented about how good the fire smelled as he got close to the island. I thought it was kind of strange that he could smell it so far away; it wasn't a very big fire. But as there was nobody else on the island, I assumed that it must be ours.

He and Lindsey and Sarah took off again in the boat. I heard them returning before I saw them so I went running down to the dock to see what all the yelling was. "Get some pans and buckets," Lindsey yelled. "There is an underground fire up at John and Priscilla's."

We took what large pans and buckets we had and went up the island about a quarter mile. I was surprised at what I saw. A fire had burned from the firepit, under the ground to the roots of a large white pine. Then it came through the ground in various places exposing the roots deep below. There was no flame but lots of smoke coming from the small holes that surrounded the base of the tree. We started hauling water from the lake and dumping bucket after bucket into the holes. Still we could see smoke. So we just kept dumping. We didn't leave until we were satisfied it was out. We figured a canoeist who thought that he was camping at the public campsite had started this one. I'm sure he thought he'd put out his fire and never dreamed it had gone underground where it had smoldered until we luckily saw it.

We had one other experience with underground fires, and this time it was on our land. We were at home in New Hampshire one day when we got a call from Dave and Jane Graham from the other side of the island. They said they had good news and bad news. The good news was that our big boat had sunk. The reason it was good news was because when they stopped to pump it out and get it floating again, they discovered an underground fire burning all around our fire pit. They had taken the pump and used it to pump water all over the area until they felt that it was safe.

We couldn't believe it. We felt we had always been so careful. Tom had put tons of sand into the bottom of the fireplace. We didn't think we had left a fire burning when we went home. But obviously, somehow a spark ignited the duff and had burned slowly and silently for a number of days. When we arrived, we found a huge hole all around the pit and a pile of nails which was all that remained of two log benches. Once again, we were lucky.

Yellow jackets are one type of wildlife that we like to avoid. They are masters at cleverly hiding their nests in the ground and finding one leads to unexpected results. My encounter was one August evening when I was at the lake by myself. I decided to do some tree trimming. I had the long tree saw, the rake, the large red clippers and a pair of small hand clippers. It was starting to get dark, but I couldn't seem to stop clipping. I was trimming branches right along the shore when I suddenly heard buzzing. It took only a moment to realize that I had found a yellow jacket nest. Why I didn't run right then, I have no idea. Instead, I kept clipping until suddenly I felt a sharp pain in my finger. "Yeow!" I yelled as I brushed a yellow jacket off. Then I felt another sting on my ankle and I realized that I needed to get out of there right then.

I ran along the rocks and out onto the dock. My mind was totally focused on what I figured would be the best way to escape the bees, which was to jump into the lake. I dropped the hand clippers, kicked off my sandals and dove into the water at the end of the dock, clothes and all. As I was hitting the water, I realized that I had not removed my glasses.

This was the third time that summer that I had jumped in with my glasses on. Once was at my sister's pool in Florida, so they were very easy to see once we stopped hunting through the house and realized where they were. The next time was here, and again I hunted all over the cabin before putting on my snorkel and finding them sitting at the bottom of the lake. This time I *knew* where they were. The problem was that by now it was pitch black and there was no way I could see into the water.

I slowly dripped my way up to the cabin and started hunting for a flashlight that would be both bright enough and waterproof. Once I found one, I dripped back down to the lake and put on my snorkel and got into the water. I had a fairly good idea of where the glasses would be so I started shining the light. I found if I held the flashlight as far down as possible, I could see the bottom. However, if I went too far down, I would fill my breathing tube with water and would have to come to the surface to empty it.

I was relieved when I finally spotted the glasses and dove down for them. There was still a problem, however. In order to dive down, I had to come up out of the water and by the time, I went down again, I had lost sight of the glasses. And then I would run out of breath and have to come back up to the surface. After a couple of futile attempts, I realized that the glasses were right in front of a large rock and if I could find the rock, I could then find the glasses. And that was what I did.

After my successful rescue, I dripped once more to the cabin, where I realized that my two yellow jacket stings were really painful. The end of my finger was swollen and my leg was red where the stinger had gone in. I started thinking about all the people who were allergic to insect stings and here I was alone on an island. I ran for our copy of Backwoods Medicine and was reassured when I read that highly sensitive people would be in trouble in minutes. So I figured that while I might be in some pain, I would live.

In recent years, Tom began to explore some realistic emergency procedures for the island in case of a serious injury. He talked with various people, did a lot of thinking and ended up writing what he calls "An Emergency Protocol." Hopefully we will never have to use it, but everyone feels better having a well thought out plan to follow.

15 ~ Island Critters

BEING IN THE WILDERNESS ONE expects to see lots of unusual wildlife. The reality is that while we occasionally see a moose or an otter, our most frequent wild visitors are red squirrels. But we are always on the lookout for any kind of animal life, whether we are on the lake, hiking in the woods or just sitting in the cabin.

The infrequency of moose visits means that it is still exciting when a one makes its way across the front of our cabin. This has happened just a few times--once was when my niece, Peggy, was sleeping outside on a platform in some trees. She looked down and there looking back at her was a large bull moose.

We sometimes see moose swim across to the island from the mainland. Once I was here alone and thought I saw two very large loons off in the distance. As they got closer, I realized I was seeing the antlers of a moose. Another time we watched a female and her calf swim across. Even when we don't see a moose, we often find their droppings along our trails and know that they are around. One early morning in September, I was awakened by a very loud sound that was somewhere between a moo and a bray. It was coming from the direction of the lake and got louder and louder and then began to slowly fade into the distance. When I got up, I discovered moose tracks in the sand all along the shore. When I told one of our island neighbors about it, he told me it was the call that the female moose makes to attract a mate. Author Bill Silliker in his <u>Maine Moose Watcher's Guide</u>, calls the early fall the Crazy Season because, not only are the females running around braying out their moans, the bulls are answering with their own, "ugghh… ugghh" and fighting each other for the cows.

When our bunkhouse was first built and was uninsulated, it was a frequent home to mice. We set traps, which would go off at unexpected times. One time some visiting honeymooners said they heard a loud snap in the middle of the night and then lay there and listened to the death throes of the mouse as he tried to free himself. He finally threw himself and the trap off the rafter onto the floor where he was picked up and thrown outside.

Tom was very excited to discover what happens to the dead mice. It was his job to check the mousetraps in the bunkhouse. One particular afternoon, he found a mouse in one, but this mouse was moving. It was sort of going up and down in rhythmic motions. While his immediate reaction was that it was dead, the motion was mystifying. As he got closer, he saw that it was not life that was moving the dead mouse; it was nature's recycling process in action. Two beetles, each about the size of a fingernail, were moving it around. He took them outside and set them in the dirt. Immediately, the beetles started moving the dirt out from under the mouse. Over the next several hours, he kept returning to watch their progress and by late afternoon the mouse was gone, buried. The next morning he checked the trap line again and found another mouse. He put it outside and almost immediately two beetles came out of the ground and started to bury this one as well. As he stood there, he noticed the ground move a bit off to his left. He carefully pulled back the dirt and there were the remains of a mouse that had been caught two weeks earlier. All that was left was the fur, tail, and skeleton. We later learned that our beetles were "Carrion Beetles." Their job is to dispose of the dead.

Although many people have seen otter, I am still waiting. One winter Diana and I were walking on the lake and found unusual markings on the snow. It looked like someone had been pulling a small toboggan, but the tracks would disappear and we'd see running paw prints. Diana, being a school librarian, did some investigating when she returned home and found that this was an otter track. They run along on the snow, and then go down on their bellies and slide for a while. We also found scat around a huge shore boulder where, given the way the ice moved and shifted, holes had been created that would give them access to the water underneath.

Our favorite animal is the loon. Their nighttime cries are part of the fabric of our lives. We often see them float by the dock, sometimes alone or in pairs. By fall they come by in groups of 15-20 as they gather together for the migration south. One year when I was teaching, I decided I would model an animal report for my fourth graders and chose the loon. I learned that they have four different cries and that each has its own purpose. The tremolo is sometimes described as insane laughter and indicates agitation or fear. The wail that we hear at night is a way for loons to keep in contact with others on the same lake. Male loons use the yodel to defend their territory, and the hoot is a quiet call that is used to keep in touch with chicks or nearby groups.

One of the best loon stories is Lenny's. The way he tells it is that he was fishing and saw that there were a couple of loons off the end of the dock. Always fearful that they might go for his lure, he waited until they had moved on and then he cast his line. He felt a tug and began to reel in his fish. When his line was about ten feet from the end of the dock, he decided that he must have a really big fish as the tug of war was becoming much more intense. When the rod bent and the line went under the dock, he pulled even harder. Suddenly up popped a loon with his fish in its mouth. He reached down and grabbed it by the neck. The loon gagged a couple of times and before he knew it, he had a fish in one hand and a loon in the other. He stared into its big red eyes before throwing it back into the lake where it dove in and disappeared. Lenny released the fish and decided that fishing was over for the day.

We have had a number of dogs during the years we have been coming to the island, and by far their favorite island animal is the red squirrel. They chase them from tree to tree, and if the dog begins to lose interest, the squirrel looks down and starts chattering, which starts the game all over again. One of our dogs even managed to catch and kill one of the squirrels. We figured it was pure luck, as most always it is the squirrel that comes out on top...literally and figuratively.

One of the surprises on the island is that we have relatively few bugs. Because we bought our land in the fall, we were not sure what would happen when black fly season came around in the spring but were expecting the worst. May arrived and there were no flies to speak off.

The same is true of mosquitoes. The mainland can be swarming with them but we have many fewer. It is the biting flies that appear later in the summer that drive us nuts. We all sit on the deck with fly swatters in hand. Kelly claims she got well over a hundred in just one afternoon.

We frequently go on wildlife expeditions around the lake. In the early evening, we'll get in the boat and head up the lake in the hopes of seeing something out for an evening meal. We have seen a number of moose this way as well as a family of geese. We have seen more and more eagles in the past few years.

I have heard coyotes on the mainland, and we are still waiting to see a bear, although we did find scat on one of our trails that looked like it could possibly be from one. We are serenaded by toads during early July and occasionally see the snake that lives under the wooden walkway. Duck families visit hoping for a handout. Each new spotting of some of the more rarely seen animals always brings a shout, and we are all excited to see what next has appeared on our island.

16 ~ Kayaking Richardson

OUR MOST ADVENTURESOME WILDLIFE EXPEDITION was the day we went to Richardson Lake. Tom and I had always wanted to explore this lake, which is the next one in the Rangeley Lake chain. We loaded the double kayak into the back of the big boat and took it down to the end of the lake, where we could then carry it through the woods to Richardson. It was a little more work than we had anticipated, but was worth the effort when we finally set off to explore a new body of water.

We headed across the water to the far shore to explore Cranberry Cove, a place we had seen on a map. As our eyes scanned the shoreline for the entrance, we noticed a tall dead pine with something sitting on one of the topmost branches. We grabbed the binoculars and saw a bird perched on it. One foot grasped the thick branch; the other seemed hooked onto a smaller twig. The only movement came from the slow turning of his white head as his eyes searched the lake. His brown body remained absolutely still. It was a bald eagle. The grandeur of this magnificent bird was apparent as he slowly surveyed the lake. We quietly paddled closer and then sat as still as he was while we watched him.

Soon we headed into the cove, which was right behind the eagle's perch. We found it nothing like the lake. From open water we entered a world of shallow waterways, marsh grasses and hidden stumps that lay waiting to catch our kayak. After getting hung up a couple of times, we decided to move back into the lake.

As we moved slowly down the shore, we watched two ospreys glide into a tree. I love these birds, especially when they catch their fish. It amazes me how they can come up with the fish in their talons always facing forward. I wished that they would dive but instead they flew off.

Next we headed further up the shore until we came to a place we could land. While I swam, Tom used the binoculars to watch a mother loon teaching her two babies to dive. We see a lot of loons on Mooselookmeguntic, but we had never before seen babies. We got back in the kayak and headed towards them trying to get a closer look without getting too close. They were testing their independence, but we noticed that mom was never far away.

We then headed back across the lake. By this time, the wind had picked up and the paddling had become harder. My arms were beginning to ache and I began to wonder if I would be able to make it back to the portage.

We finally got back to the landing. I wanted to whine about how tired I was, but knew it wouldn't do any good. We still had to get the kayak back to our lake. So instead, we picked it up, walked a few feet… then set it down. Picked it up, walked a few feet…set it down. By the time we got the kayak back to Mooselook and the boat, we were ready for the cabin and a long nap.

17 ~ Having Fun at 20 Below

WINTER IS ONE OF MY favorite times to come to the island. What was open water in the summer becomes a huge white field in the winter. Now that the road is plowed all the way to the boat launch, it is an easy hike across pulling our sleds with our gear. Of course, forty mile an hour winds add to the challenge, as do frigid temperatures. Once here, we unlock the cabin only to find that it is colder inside than out! We get the woodstove going and it quickly heats up. Then Tom can go out with his ice auger and cut a hole in the ice for water. Before we had the auger, we tried other ways of cutting into ice that can be two feet thick. One of the first was a chain saw. It seemed to work great until we tasted the water. It was tainted with the flavor of chain saw oil that refused to go away. Next a long drill bit was used. Before he drilled, Tom would take an ax to chip out a large collecting bowl in the ice. Then he'd drill down until he hit water. The water would bubble up and fill the bowl. This is basically the same principle we use today--but the ice auger has replaced the drill.

While winter is a great time for x-country skiing and snowshoeing, there are three key jobs that we like to get done on these trips. It is the time Tom goes into the woods and marks new trails. This job has become less important as he runs out of places to put them.

One job that will never end is the winter burn! This has become a tradition. All summer we cut up trees that have fallen and clear out brush. All of this debris goes into one giant pile that is then covered over with a tarp to keep it clear of snow. It is usually Martin Luther King's birthday weekend that Tom, Lenny and LeRoy come up and set it alight. Some years the burn is truly magnificent…flames shooting up into the sky while the logs below burn red-hot.

The third job that has seemed to materialize over the few years is bringing the firewood down from the back acreage. While there are certainly plenty of trees near the cabin, they are all either pine or birch, both of which burn very quickly in the stove. The treasured maple is a quarter of a mile up the hill over one of Tom's meandering trails. But he and Lenny have decided that it is needed for burning in the stove at night. Their procedure is to snowshoe back to the pile of logs that has been cut and stacked during the summer. They tow sleds, which they pile with wood, and then head down the path back to the cabin. It is hard work, but truly appreciated during the cold winter nights.

Tom, Lenny and LeRoy hauling down wood

One of the first winter trips involved Tom, his brother Dave and Lindsey who was about 12 at the time. This was before we had anything out at our site, so not only did they have to bring what they needed--tent, food, sleeping bags etc., they also decided to haul over a large wooden box that we could use to leave gear in between visits. They assumed that sliding it over on the ice would be easy. It wasn't--as Lindsey relates in her version of the trip.

I could not believe it. I was on my way to Maine for my first winter camping trip and was extremely excited and nervous. I was nervous because it was so cold and windy and I did not know if I had brought enough clothes to keep warm. I was excited because I knew I was going to have an awesome time and, it was my first winter trip.

We got to the lake about 6 p.m. and it was dark out. The plan was to walk over to the island pulling a huge box on a toboggan. The box was going on a one-way trip. It was going to be left there as a place to store our equipment. We started out with all of our supplies in it, except for my backpack that I carried.

We got tired because the depth of the snow made walking difficult. We finally got so exhausted we left the box on the ice and took the tent, clothes, pads and our sleeping bags to the land. When we finally got there from our hard trip across the lake (I complained the whole way), we put up the tent and threw in the sleeping bags and pads. Dave and Dad left to get some more of the supplies while I laid out the sleeping bags and pads. Then I lay down in my sleeping bag--which felt very warm and cozy, and started to think about the rest of the weekend ahead of me. However, I was so tired and comfortable, I soon fell asleep. Then the guys woke me up but not before they had chopped a hole through the ice so we could get water. Since the ice was so thick, they finally gave up on the axe and used the chain saw. SMART! So the water tasted like oil. It was outrageously sickening at first, but then you got used to it. In the meals, you could not taste it.

The tent in winter.

That night after dinner, I went to bed. Dave and Dad stayed up later. We had to put the boots and clothes that we were going to wear the next day in our sleeping bags so they would not freeze (which mine did anyway). Feeling like an elephant in the North Pole, even with two sleeping bags, I finally fell asleep.

The next morning I peeked out of the tent, having no idea what time it was, and saw that Dave and Dad were already up. So I decided to get dressed and go outside. We all got breakfast ready and ate it fast, because it got cold very quickly. Oatmeal is disgusting when it is cold.

Unfortunately, after breakfast, we had to go and get the box. The trip to the box was not so tiring, because Dad and I only had to pull the toboggan. It was easy until we got to the deep snow, which drifted to a height somewhere between my ankles and knees.

Anyway, we got to the box just in time because if I had taken one more step, I would have collapsed. At least half of the way we had to walk through the deep snow. But I fell over anyway. We got the box on the toboggan and tied it on.

We headed back going at a steady pace until we got to a point where the box was dragging in the snow. So we had to stop and fix it. Then I would get at the back of the box and on the count of three I would yell, "MUSH!" I would give it a hard push to help them get started again. This happened every time they would stop, which we had to do a lot, because the box kept on dragging in the snow.

We arrived back at our land about noon—or our stomachs thought it was about noon—so we ate our lunch. Then we finally got to do the other thing that we hoped to do. This was to burn wood that was in the way, so we could eventually build a cabin and the land would not look so much like my room. We started to burn and then we burned and burned and burned. We kept the fire going the whole afternoon. During this time we took turns laying out on the toboggan and dozing off. When it was my turn I fell asleep and got sunburned--and the temperature was only 40 degrees.

That night I fell right to sleep, but we all woke up several times because the wind was blowing so hard the tent was moving and you could hear it.

Finally morning came. We had to eat our breakfast even faster because it was ten below zero. The temperature had dropped fifty degrees overnight. My boots were frozen and I mean, "frozen." I literally could not get my feet into my boot, but finally I pried my feet into them. We got everything together and put it all on the toboggan, except of course for the box. I was *so excited because I would be able to have some h-h-heat when we got back to the truck.*

It was pretty easy on the way back to the truck because snowmobiles had been zooming around yesterday, so now we had something for the toboggan to glide on. As soon as we reached the truck, we unloaded our gear, hopped in and turned on the truck and heat. I was so excited to be heading home. On the way we stopped at McDonald's. I felt like a freak because I had not combed my hair in two days. I smelled and I had the strangest clothes on. We talked about how much fun we had had, even though it was cold. I am willing to go again as long as I bring my warmest clothes!

Winter is also a time to play. There is nothing like snowshoeing through the woods on a bright sunny day. No matter how cold the temperature may be, walking on snowshoes creates an internal furnace that soon has the sweat dripping down our backs. There are mice prints dotting the snow and occasional rabbit tracks that lead us off into the brush as we try and follow their path. One year we came upon an area where moose had been bedding down. And we always hear the chickadees singing in the trees.

One of our more memorable late February hikes was with Holly and Lenny as we tried to find "something bright red" that Tom had spied when he'd come early to warm up the cabin and haul wood. Back on the trails we headed until he was sure we were at the correct spot. Nothing could be seen from the trail, so into the woods we went, searching in circles for this mysterious red thing. The snow was deep—probably two feet, so we were falling often. Finally Lenny gave a yell and reached into a clump of bushes to pull out a deflated Mylar balloon. It was heart-shaped and said, "I love you" on it, obviously a leftover from Valentine's Day. It now hangs on one of the walls in the cabin, another found treasure.

Cross-country skiing is something we enjoy, but find more problematic. The back trails are too uneven and winding to make skiing possible. The lake conditions are erratic. Because of the wind, the surface can be swept clean in places and drifted in others. When conditions are right, however, skiing the lake is exhilarating. Going across the lake to the unplowed South Arm Road can also make for good skiing.

Peggy Herbert

There are those who love the winter at the lake and the different challenges it provides. And then there are others who would rather stay snug and warm in New Hampshire.

18 ~ The Next Generation

Emma, Cooper and Regan

NOTHING COULD HAVE PREPARED US for how cabin life would change once we had grandchildren. Having young children at the lake is an experience that we missed with our own daughters, as they were 11 and 14 when we began our adventures on the island, and consequently they did not grow up with the cabin being a regular part of their lives.

Emma was born in 2003, Cooper in 2005 and Regan in 2007. They all started coming as infants and love it in a way that you only can when you are a child. They begin asking about when they can go again almost as soon as the summer ends. When I recently asked Emma how long she thought she could stay before she got bored, she thought for a moment and then said, "A year." Kelly later told me that when they got home she announced that she had changed her mind. A year was not enough time. She decided she'd like to live up at the cabin forever.

Of course there have had to be modifications made in our regular routines. Jumping in the cold lake at six in the morning is not a viable option for infants and toddlers, so we had to come up with a way for them to take baths. We have found two successful methods. When it is warm out, we have a plastic kiddy pool that we put on the deck and fill with warm water. While the kids splash around, an adult tries to get them washed. This usually means that whoever that person is, she or he will end up wet as well.

Cooper in the cooler getting a bath

On colder days, we hit upon using one of our coolers as a bathtub. This is the one-at-a-time bath, but the kids think it is great and are willing participants. As they grow taller, however, they will outgrow this inside bath and will have to use the lake as the rest of us do.

Using the bathroom has been another big adjustment. First of course were all the diapers. We found we were making extra trips to the Rangeley Plantation dump just to get rid of them. Emma graduated to the children's potty-chair as soon as she was trained. We kept this in the small downstairs room and made frequent trips to empty it. Even though she was encouraged to use the outhouse, she would have none of it. The few times she went up the hill with one of us, she stopped cold at the door and refused to enter. The summer she was almost four was a turning point. She would not only use the outhouse, she'd run up the hill on her own needing no adult to hold her hand. This was a big relief to everyone. Cooper did not show the same hesitancies about the outhouse, but does like someone to go with him and wait. Regan was actually trained at the cabin. It was the perfect place to run around with just a t-shirt, and it only took a couple of days before she too was wearing "big girl underpants." At least for now, no more diapers!

The summer that Emma was four and Cooper two was the year that they "took to the woods." They would put on their backpacks, get snacks and head out on the trails with Grandma and Grandpa and Libby the dog. We'd stop for a snack, and after eating, they'd take out their notebooks and pretend to take notes of what they saw.

As they have gotten older, they have become more independent, following trails on their own and then dragging some adult back to see something really interesting like "monster tree" or dog poop. We also find them making "fairy houses," the famous natural creations in the forest of Monhegan Island off the coast of Maine. I showed them how to make Inuksuks, Inuit rock cairns that were made in the arctic to show direction or to honor a special place or person. They have gone from just piling up rocks to beginning to make simple people. While Cooper and Emma are busy doing all these activities, Regan can be found wandering around happily picking up sticks or playing with the dog.

Peggy Herbert

Cooper and Emma taking notes and eating M&M's.

Swimming lessons at the YMCA have given them more confidence in the water. The first time Emma and Cooper swam out to the swim raft with their life jackets on was momentous. We took pictures and cheers echoed off the mountains. Now they want to do this before the water has even warmed up---although usually a short swim in frigid temperatures has them rethinking the idea. All three of them can spend hours in the water, diving, going on the tube and running and diving off the end of the dock. Emma, at almost seven, surprised us all by getting up on water skies on her third try and then went zipping around the lake.

One of the favorite activities is swinging. Tom and his brother Pete put in a swing between two tall trees. It has a wooden seat as well as loops to attach a plastic child swing. Emma and Cooper learned that one of them could sit on the seat while the other sat in the toddler swing and they could both yell, "Higher, we want to go higher!" It is Regan who can swing endlessly. With an adult to keep her moving, she is happy as can be. When winter winds blew down the trees that supported their swing, Tom immediately found two others and before long, the swing was back in action.

I wish that I had had a tape recorder the night that Emma, Cooper and I took a quilt down to the dock and laid on our backs to watch the stars come out. Their excitement grew and they began to see more and more stars and then an airplane, and not long after that, a satellite appeared. Soon the questions started. "Where do all these stars come from?" "Where are they in the day?" "Is our sun the biggest star?" And then Cooper asked, "Where do we come from?" Emma quickly jumped up and announced that she could answer that. "Well, Cooper. Mom and Dad told me this and I didn't believe it at first but now that I am older I do." (I was holding my breath wondering where this was going.) "Cooper, we come from monkeys! You see one monkey had a baby with less hair and then another monkey had one with even less hair until here we are." The magic of that evening stays with me as I remember listening to these two asking the questions that many of us ponder still.

It always amazes me how happy children can be with some water, sand, a few shovels and lots of space to run around. In 2006, Richard Louv wrote *Last Child in the Woods: Saving Our Children from Nature-deficit Disorder.* He feels that direct exposure to nature is essential to healthy childhood development and for the physical and emotional health of children. I am glad to think that Emma, Cooper and Regan will have plenty of opportunity to explore the out-of-doors and will experience nature in such an extraordinary environment.

19 ~ Stories from the Grandchildren

I HAVE ENCOURAGED THE GRANDKIDS to write, (or tell) stories about lake adventures. I have always found that there is nothing like seeing your own words in print to encourage both reading and writing. Here are some stories that Regan, Cooper and Emma have written

Regan and the Duck (age 3)

I was in town. We were eating ice cream. I had peppermint stick because it was pink. There were ducks there. I threw a piece of my cone on the ground. A duck ate it. Then I put my hand in my lap and a duck bit my finger. He wanted the rest of my ice cream cone. He nibbled my finger a little bit. I screamed and cried a little bit. Mom looked at me and gave me a hug. I threw the rest of the ice cream in the garbage and we left.

Regan is a Slug Wrangler (age 3)

One day at the cabin, we decided to go to Bald Mountain. I wanted to go along. I wanted to hike all the way up. I hiked along and I saw a slug. I picked it up and put it on the side of the path so it wouldn't get stepped on and dead. I kept hiking along and I saw more slugs. I picked them up and put them on the side of the path. My hands turned orange from so many slugs. Dad said I was a good slug wrangler. I just kept picking up slugs all the way to the top of Bald Mountain.

Cooper Jumps (age 3)

I was at the cabin. I played in the sand and the water. I wanted to jump off the dock. I wanted to jump to Grandma. So I stood on the dock with my boat coat and Grandma stood in the water. I jumped and Grandma caught me and swung me through the water. I laughed and said, "Again, Grandma!" So I jumped again and again.

Swimming to the Swim Raft (Cooper, age 4)

At the cabin, I watched Emma jump off the dock and swim to the swim raft. I was four and I wanted to swim to it too. One morning, Grandma, Emma and I decided to swim out to the raft. I put on my boat coat and bathing suit and got in the water. Grandma and Emma got in too. We started to swim. It was exciting because I swam all the way there. When we got there, we climbed up the ladder and sat on the raft for a while. Then I swam back next to Grandma. I liked to spit water at her while we swam. Mom was clapping and Grandpa was jumping up and down and cheering. When I got out of the water, I was cold, proud and happy.

Emma's Kayak Adventure (age 4 ½)

When I was almost five, I went kayaking on Lake Mooselookmeguntic with Mom, Dad, Cooper and Grandpa. It was my first time on a big kayak trip. Grandpa was showing me how to paddle. We paddled past the cabin. I was getting tired so we tied a rope from Grandpa's kayak to my kayak so he could pull me. My kayak was getting wobbly. All of a sudden, I felt my kayak start to tip and I fell into the water! My head went under and then I popped up. I thought, "What happened?"

My kayak was upside down and everyone was looking at me. Grandpa asked me to hold onto his kayak and he took me over to a dock on the shore. He pulled my kayak behind him while Mom and Dad followed and picked up my paddle. While he got out onto the dock, I stayed in the water and held onto Mom and Dad's kayak. I was shivering because I was so cold. Grandpa pulled me up onto the dock and put me back in my kayak. Then he pulled me back to the cabin.

Water Skiing (Emma, age 7)

One day at Mooselookmeguntic I wanted to water ski. Everyone got in the boat except my dad and me. We swam out to the boat. The boat stopped and I got the skis on. The boat started again. 3....2....1! I fell sideways and the boat stopped. I straightened myself out again. I fell again. This time I was sure I could make it. I did it! I was up! I went in a circle. And then after a few minutes, I looked at my skis. I thought I would fall. I never did. And then I was done water skiing. I could water ski!

20 ~ The End…or Just More Beginnings?

I HAVE TRIED TO SHARE some of our life here on Toothaker Island. It is certainly not a life for everyone. Some people continue to think we are crazy, but there are just as many who envy us our idyllic spot. We have learned a lot over the years, both in skills and in appreciation for what we have been allowed to do. As you can see, we are never really done. There are always new projects and new lessons to learn.

As time goes on, there will be new adventures, new people and new challenges to face. We feel lucky to have been able to create such a blissful retreat in a world where simple values are often lost. We hope we can hold on to some of these as well as pass them on to future generations.

Peggy Herbert

The cabin today.

Works Cited

Barker, Fred C.. *Lake and forest as I have known them*. Author's ed. Boston: Lothrup, Lee & Shepard, 1903.

Earle, Alice Morse. "Colonial Neighborliness." In *Home life in colonial days*. New York: Macmillan Co., 1898. 416.

Ellis, Edward. *Chronological history of the rangeley lakes region, maine*. S.l.: Sandpiper Press, 1983.

Farrar, Charles Alden John. *Farrar's illustrated guide book, to Rangeley, Richardson, Kennebago, Umbagog, and Parmachenee lakes, Dixville notch, and Andover, Me., and vicinity With a new and correct map of the lake region ....* Boston: Lee and Shepard;, 1877.

Hutchinson, Doug, and Louise M. Korol. *The Rumford Falls & Rangeley Lakes Railroad*. Dixfield, Me.: Partridge Lane Publications, 1989.

Palmer, R. Donald. *Rangeley Lakes Region*. Portsmouth, NH: Arcadia, 2004.

Perkins, Jack. "Shelton Noyes Evolves Unique Rangely Wild Land Resort Plan." *Lewiston Journal*, November 11, 1961, sec. Magazine.

Silliker, Bill. *Maine moose watcher's guide*. So. Berwick, Me.: R.L. Lemke Corp., 1999.

Chicago formatting by BibMe.org.